Measuring Cybersecurity and Cyber Resiliency

DON SNYDER, LAUREN A. MAYER, GUY WEICHENBERG, DANIELLE C. TARRAF, BERNARD FOX,
MYRON HURA, SUZANNE GENC, JONATHAN WILLIAM WELBURN

Prepared for the United States Air Force
Approved for public release; distribution unlimited

For more information on this publication, visit www.rand.org/t/RR2703

Library of Congress Cataloging-in-Publication Data is available for this publication.
ISBN: 978-1-9774-0437-4

Published by the RAND Corporation, Santa Monica, Calif.

Preface

This report presents the results of research sponsored by the Commander of the Air Force Life Cycle Management Center to develop measures of effectiveness for cybersecurity and cyber resiliency in U.S. Air Force missions and weapon systems. These metrics are developed to be suitable for informing acquisition decisions during all stages of weapon systems' life cycles. The work was conducted within the Resource Management Program of RAND Project AIR FORCE. It should be of interest to the cybersecurity, acquisition, test, inspection, and operational communities.

RAND Project AIR FORCE

RAND Project AIR FORCE (PAF), a division of the RAND Corporation, is the U.S. Air Force's federally funded research and development center for studies and analyses. PAF provides the Air Force with independent analyses of policy alternatives affecting the development, employment, combat readiness, and support of current and future air, space, and cyber forces. Research is conducted in four programs: Force Modernization and Employment; Manpower, Personnel, and Training; Resource Management; and Strategy and Doctrine. The research reported here was prepared under contract FA7014-16-D-1000.

Additional information about PAF is available on our website:

www.rand.org/paf

This report documents work originally shared with the U.S. Air Force on September 26, 2016. The draft report, issued in September 2016, was reviewed by formal peer reviewers and U.S. Air Force subject-matter experts.

Contents

Figures

Tables

Summary

This report presents a framework for the development of metrics—and a method for scoring them—that indicates how well a weapon system or mission is expected to perform in a cyber-contested environment. There are two groups of cyber metrics: working-level metrics that aim to counter an adversary's cyber operations and institutional-level metrics that aim to capture any cyber-related organizational deficiencies.[1]

The cyber environment is dynamic and complex, the threat is ubiquitous (in peacetime and wartime, deployed and at home), and no set of underlying "laws of nature" govern the cyber realm. A fruitful approach is to define cyber metrics in the context of a two-player cyber game between Red (the attacking side) and Blue (the side trying to ensure a mission). Red's strategy and tactics will be shaped by its assessment of Blue's posture and weaknesses. Likewise, Blue's posture will be shaped by an expectation of what threats Red poses. Both will continually evolve. No forethought by Blue, no matter how carefully done, will suffice in anticipating all of the possible moves Red might take in the future. Blue will need to use "static" countermeasures based on known best practices (cybersecurity), as well as adaptive, "dynamic" actions to respond to Red in real time (cyber resiliency). Both of these dimensions of cyber metrics need to span nearly the entire scope of the enterprise to capture the full range of concerns.

To measure how survivable and effective a mission or system can be in a cyber-contested environment, we must understand how well Red cyber operations are being countered. Therefore, the focus of cyber metrics must be on Red's estimated success or failure, not on the specific countermeasures that Blue might try. Blue countermeasures are important, of course, but their importance is as a means to an end—that of hindering or thwarting Red. Cyber metrics that concentrate on the compliance with lists of candidate Blue countermeasures fail to indicate whether those measures are effective, properly implemented, or sufficiently comprehensive to thwart Red. The compliance approach, therefore, is insufficient. In addition, a simple set of metrics in the form of a "dashboard" does not exist that satisfies these needs.

Working-Level Cyber Metrics

Given its complexities, a very wide range of cyber metrics is needed to monitor activities at the working level. A framework for these metrics based on a cyberattack path, motivated by the

[1] We use the term *working level* in this report to refer to the level in the enterprise of program offices and operational wings. By *institutional level*, we mean the level of the entire enterprise, including overall policies and culture and how people and organizations interact in the carrying out of their duties.

two-player game paradigm, provides both a means to ensure that the metrics are comprehensive and some context for how much is enough.

Framework and Metrics

We adopted a framework for working-level cyber metrics that focused on stopping Red's cyber operations. To be successful in cyberattack, cyber exfiltration, or via an insider, Red must execute a cyberattack path. At the most fundamental level, that attack path includes getting access to a target system, obtaining enough information about the target to effect the attack, developing or obtaining the capabilities required to execute the attack, and identifying a target useful to attack because it plays an important role in executing peacetime or wartime missions. These four activities do not necessarily need to happen in that order and might overlap in time.

Each of these four can be further divided into higher levels of indenture. For example, the need to access a system can be achieved via the supply chain, data pathways into the system, and through insiders. Each of these can be further subdivided. For example, the supply chain can be infiltrated during design, manufacturing, shipment and storage, maintenance, and disposal. A full decomposition of Red attack elements describes what needs to be measured—the ability of Blue to counter each of these Red actions.

Blue can respond to Red's actions via a number of countermeasures so numerous and complex that it is futile to try to list all possible countermeasures. These Blue countermeasures are best developed at the working level where most of the technical knowledge lies. Any lists of potential countermeasures, such as controls suggested by the National Institute of Standards and Technology, should be viewed as guidance.[2] They should supplement the working-level efforts to counter Red and not be viewed as substitutes for thought.

Focusing on the Red actions to counter provides a natural framework for ensuring that working-level cyber metrics are sufficiently comprehensive (i.e., whether all potential Red actions are covered) and knowing how much is enough (i.e., the minimal set of Red actions that need to be defeated to thwart Red). One example of the latter is in the Boolean structure of the breakdown of Red's attack path.

Red needs to have access, knowledge, capability, and impact. These four activities are related by Boolean *and* statements. Defeating any one of those four thwarts Red. In general, for Red activities related by a Boolean *and* statement, from Blue's perspective, the overall countermeasures are as good as the *best* that Blue is doing between the pair. On the other hand, Red can gain access to systems via the supply chain, standard data paths, or via an insider. These activities are connected by Boolean *or* statements. Blue must thwart all of these to be successful. In general, for Red activities related by a Boolean *or* statement, from Blue's perspective, the overall countermeasures are as good as the *worst* that Blue is doing between the pair. For

---

[2] See, for example, National Institute of Standards and Technology, 2015, and Committee on National Security Systems, 2014.

example, within the knowledge part of the attack path, Red must get the requisite knowledge, and that knowledge must remain current. Therefore, poor performance by Blue in stopping Red from getting the requisite knowledge can be partially offset by making that knowledge ephemeral (e.g., by changing system configuration, modernization).

Assessments

Assessments can be done at three levels: (1) self-assessments, primarily by program offices and operational wings; (2) validation and verification of those assessments by outside parties, primarily program executive officers, authorizing officials, and the testing community; and (3) at the mission level, by combining the assessments at the system level.

The most important product of the assessment at the working level would be artifacts that document what is being done to counter each of the Red actions. It is at this level where the technical and operational details lie. This level must do the most-detailed assessment to document the Blue countermeasures and their qualities against Red. The leadership to which the working level reports will, however, need to see some aggregation in the form of a scoring of these assessments. This aggregation is necessary to get a broader view of the mission without needing to review all of the artifacts produced at the working level. We recommend a maturity index for this scoring, as described in Table S.1.

Scoring of the maturity index for each Red vector should be done conservatively. To progress upward to higher maturity scores, issues at each maturity level must be addressed as completely as feasible. One weakness in Blue's countermeasures against a Red vector can be all that Red needs to prevail. Therefore, even if some countermeasures against a Red vector have been identified, implemented, tested, and found adequate, the response to this Red vector remains immature if others have not been properly identified. For example, if countermeasures for accessing a weapon system through communications paths have been comprehensively identified, implemented, tested, and found adequate, yet all the paths through which Red might gain access via subsystems remain to be identified, the scoring for Red access via data pathways remains "most immature."

Table S.1. Maturity Levels for Working-Level Metrics

Level	Maturity	Characteristics
1	Highest maturity	Solutions to counter a Red vector are tested, exercised, and found to be adequate.
2	Mature	Solutions to counter a Red vector are implemented with continuous monitoring.
3	Intermediate	Solutions to counter a Red vector are identified.
4	Immature	How Red might act by a vector is understood in the context of the system or mission under review, and a baseline trusted state is defined.
5	Most immature	Awareness of how Red might act by a vector against a system or mission is

Level	Maturity	Characteristics
		inadequate. Such examples for access include incomplete knowledge of standard pathways for data inputs and outputs of a system.

Institutional-Level Cyber Metrics

The overall status of cybersecurity and cyber resiliency is not just a sum of the assessments of all of the parts of an enterprise. It emerges from how all of those parts interact. So, even when adequate working-level metrics are in place, deficiencies at the institutional level could still go undetected. Several well-studied cases indicate that institutional deficiencies of organizations can play a large role in failing to carry out a mission.

Lessons from failures of organizations indicate that deficiencies are generally known somewhere in the organization before the failure. The information was not channeled to the right authorities and was not properly assessed for risk. A large body of literature has concluded that major institutional malfunctions occur when fluctuations in the state of the enterprise exceed its ability to adapt to its environment—in other words, when fluctuations exceed the organization's resiliency. The gradual erosion of the margin of resiliency is a phenomenon called *drift*. A key indicator of drift is when senior leaders' perceptions of how operations occur differ from how operations really occur.

Organizations that successfully avoid catastrophic failures reduce drift by collecting information from all members of the organization, triage it, assess it to create meaning, and channel key information to senior leaders outside the normal chains of command. Such a process can be assessed via a maturity index for institutional-level issues, as outlined in Table S.2.

Table S.2. Maturity Levels for Institutional-Level Metrics

Level	Maturity	Characteristics
1	Highest maturity	Lessons are documented and distributed; processes are reviewed periodically for improvement.
2	Mature	Recommendations flow to relevant decisionmakers.
3	Intermediate	Concerns are assessed by independent subject-matter experts and meaning is extracted from the reports in the form of recommendations.
4	Immature	Formal mechanisms capturing key elements of the Aviation Safety Reporting System, including confidentiality, exist to report concerns regarding cybersecurity and cyber resiliency.
5	Most immature	No mechanisms exist outside normal reporting through a chain of command for concerns that individuals have regarding cybersecurity and cyber resiliency.

Implementation Issues

The implementation of cyber metrics presents challenges, both at the working level and institutional level. To varying degrees, decisionmakers at all levels will need insights from working-level cyber metrics. We caution, however, that the most senior leaders must delegate decisions to where the locus of information lies. Most of the technical information about selecting and assessing Blue countermeasures lie at the working level. Senior leaders must focus above the technical level, looking for systemic issues at the working level and institutional-level deficiencies.

All decisionmakers must grapple with the inherent uncertainties in cyber metrics. They need to resist the temptation to press for inappropriate levels of precision and stability for working-level metrics. They must also foster a culture of risk management. That means that although some Blue countermeasures should have high maturity index scores if the Red threat is high and the consequences of not countering it are also high, many should have low maturity indexes when the Red threat is low, the consequences of failure are less dire, or when risk can be accepted because weak performance in one area is compensated by stronger performance in another. Resources are always limited, and when managing (rather than eliminating) risk, scoring low in selected areas is acceptable.

Finally, sound metrics require good measurers. For cyber metrics, virtually every corner of the enterprise plays a role and, therefore, the measuring scope is vast. On top of that, the Red-Blue cat-and-mouse cyber game is fluid and requires deep reflection and insight. Therefore, Blue personnel are the most critical resource in cyber monitoring. All personnel need to participate, and all need some level of training and skills. Because the observations are so often qualitative rather than quantitative, personnel must also *communicate* rather than just *report*. In many ways, defining the right cyber metrics is the easier, first step. The hard, next step is hiring, training, retaining, and keeping current a skilled workforce to execute those metrics. Well-specified cyber metrics are only as good as the personnel who report them.

Acknowledgments

We thank Lt Gen John T. Thompson for initiating the project and Lt Gen Robert D. McMurry, Jr., for continuing the sponsorship of the project. In the U.S. Air Force, we especially thank Dennis Miller, Jeff Stanley, Eileen Bjorkman, Mitch Miller, and Danny Holtzman. Each provided key support and helpful discussions and critiques. Several others in the U.S. Air Force helped us in many ways. They are too numerous to acknowledge individually. We also thank a number of individuals in various organizations in the U.S. Navy, the National Security Agency, and the Defense Advanced Research Projects Agency for their helpful discussions.

At RAND, we benefited from conversations and critiques from Mahyar A. Amouzegar, Lauren Kendrick, Lara Schmidt, and Laura Werber. We thank Quentin Hodgson and Forrest Morgan for their constructive official reviews and improvements to the report.

That we received help and insights from those acknowledged above should not be taken to imply that they concur with the views expressed in this report. We alone are responsible for the content, including any errors or oversights.

Abbreviations

APA	additional performance attributes
ASRS	Aviation Safety Reporting System
FBI	Federal Bureau of Investigation
FISINT	foreign instrumentation signals intelligence
JCIDS	Joint Capabilities Integration and Development System
KPP	key performance parameter
MOE	measure of effectiveness
MOP	measure of performance
NASA	National Aeronautics and Space Administration
NIST	National Institute of Standards and Technology
OODA	observe, orient, decide, and act
PEO	program executive officer

1. Developing a Framework for Cyber Metrics

> "Objective evidence and certitude are doubtless very fine ideals to play with, but where on this moon-lit and dream-visited planet are they found?"
>
> — William James[1]

The purpose of this report is to present a framework for the development of metrics—and a method for scoring them—that indicates how well a weapon system or mission is expected to perform in a cyber-contested environment. The results are in the form of working-level metrics and institutional-level metrics that aim to capture the full range of concerns.[2] Despite this scope, this framework and associated metrics will not be the last word on the topic. Developing metrics for military superiority has been a long-sought goal, and this elusive quest is yet more challenging for the cyber environment. We argue in this report that any cyber metrics will need to be continuously re-evaluated and refined.

The only way to truly know how well a system or mission will perform when under assault in cyberspace is to observe it when under attack. Waiting for an adversary to attack a system to see how well it holds up is not an option. It is also not practically possible to simulate all possible attacks or all possible states of a system at the onset of the attack. Nevertheless, decisionmakers need to have information on survivability and effectiveness prior to an actual attack.

Yet, without accurate feedback on the performance of a system, mission, or organization, leaders and other decisionmakers throughout an enterprise are left partially blind. Without sufficient insight into how well operations are working, they cannot make sound decisions or learn about the effects any previous decisions might have had. Even armed with a set of sound metrics, leaders are still insufficiently served if those metrics leave key dimensions unobserved. Not only are those gaps blind spots, areas of salient feedback can unconsciously focus the goals of an organization on the issues of those areas. Domains that are not measured by an organization are often ignored by decisionmakers.[3]

Outside of the cyber domain, over time various metrics have been developed and implemented within the Air Force. Many of these metrics, and the processes used to develop new metrics, are products of the organizations and cultures of both the acquisition and operational communities that evolved to deal mainly with kinetic attacks, to which they are well adapted.

---

[1] James, 1896, p. 336.

[2] We use the term *working level* in this report to refer to the level in the enterprise of program offices and operational wings. By *institutional level*, we mean the level of the entire enterprise, including overall policies and culture and how people and organizations interact in the carrying out of their duties.

[3] Arrow, 1974; Jensen and Meckling, 1992.

Nonetheless, some of these metrics and processes are also useful in understanding mission assurance in a contemporary cyber environment. However, we argue in this report that the current cyber environment differs in some fundamental ways from an environment characterized by kinetic attacks. These differences call for a fresh look at what a comprehensive set of metrics might look like for the survivability and effectiveness of weapon systems in a cyber-contested environment.

In this report, we present a framework for measuring the cybersecurity[4] and cyber resiliency[5] of weapon systems and an associated set of metrics. The framework helps, in part, to reveal where strengths in one area might partially offset weaknesses in another. We further discuss how those metrics can be scored in ways that are useful for supporting decisions. The metrics are aimed at supporting program offices and authorizing officials in risk management and in defining requirements, both operational requirements and the more-detailed requirements for system design used in contracts, the latter often referred to as *derived requirements*.

The Purpose of Metrics

Measures and metrics exist to support decisions. What kinds of metrics are needed, how often they ought to be collected, what level of fidelity they should possess, and in what context they should be framed and presented depend on the decisions the metrics are meant to inform.

Regardless of who the decisionmaker is—whether it be a systems engineer in a program office or the Chief of Staff of the Air Force—all actors need to be guided by a common set of goals coherent across the enterprise (goals shaped by a shared understanding of the outcome that the enterprise seeks). The goals indicate what needs to be done, which guides what courses of action ought to be under consideration by decisionmakers, which, in turn, determine what needs to be known (information that must be measured and disseminated). The logic runs in that order, not in the reverse.[6] It is important to define metrics based on accepted goals because, once metrics are in place, it is a natural tendency for those metrics to drive organizational goals.[7]

For our purposes, the goal is to achieve an acceptable level of cybersecurity and cyber resiliency for mission assurance, and the desired outcome is to make the task of an adversary conducting cyber operations against weapon systems difficult and to minimize the impact of any cyberattack on operational missions. It is an exercise in continual risk management. The

---

[4] The word *cybersecurity* has had a variety of definitions in the Department of Defense over the past several years. For the purposes of this report, we will use the term to refer to actions, largely defensive, meant to provide assurance of confidentiality, integrity, and availability of cyber systems.

[5] The term *resiliency* is much in vogue in the Department of Defense and several meanings have circulated. In this report, we use the term *cyber resiliency* to mean the ability of a mission to absorb and to recover from a cyber operation while retaining an acceptable level of mission operations.

[6] See Snyder et al., 2015, pp. 13–16; and Office of Aerospace Studies, 2014.

[7] Jensen and Meckling, 1992.

principal groups of decisionmakers that we aim to support with the metrics described in this report are: the acquisition community, the requirements generation communities, the authorizing officials, operational squadrons and wings, testers, and senior leaders.

These various groups have different authorities and therefore require different types of information to make decisions. Those doing systems security engineering in a program office and writing derived requirements for a contractor need feedback at a detailed, technical level regarding system configuration. Those at the level of a program manager need metrics that span all aspects of the program. That set of metrics should help guide where to allocate resources and accept risk across the program. And, leaders above the program level need metrics that extend beyond a single program, metrics comprehensive enough to give a full picture of the security and resiliency of a mission.

Some metrics, therefore, need to provide the specificity and detail to support decisionmaking at the working levels of the organization. However, other metrics (or a combination of metrics) are needed to provide a useful enterprise view and to guide and hold accountable lower-level members of the organization. This focus on the consumers of metrics leads us to some desired attributes of metrics and an integrated framework.

Decisionmakers need metrics that comprehensively cover the full range of decisions they need to make. Gaps will exist in working-level metrics of cybersecurity and cyber resiliency reflecting the disparate authorities and responsibilities of program managers, authorizing officials, or other decisionmakers. Institutional-level cyber metrics should close as many of those gaps as possible. Not everything of interest is easily measured, so some gaps will persist. An area for which there are no metrics is an area that is likely to be ignored.[8] The existence of gaps and their substance is something worthy of measuring in its own right.

Alignment of metrics with decision authorities and comprehensiveness for decisionmakers are ideal goals to strive for. However, roles and responsibilities are allocated within an organization for many competing reasons, and that can lead to overlap and gaps in authorities relative to metrics of how well adversary cyber operations are mitigated. For some things the defender might want to do, there might not be a single person or organization in charge. Also, limitations in the ability to monitor a desired attribute of a system will inevitably leave some gaps in feedback relative to the ideal.

To be useful for making decisions, metrics must be defined in a way that can be collected and assessed without undue resource expenditure and in time to be useful for decisionmaking. Metrics should be defined such that the needed data are reasonably easy to collect and their scoring is repeatable. Yet, the metrics must also have high enough fidelity to support decisionmaking, which implies that uncertainties and limitations of the metrics should be assessed and reported. That applies to both the metrics themselves and the values or scores given to them.

---

[8] Arrow, 1974, pp. 49–53.

Some decisions, such as systems engineering decisions during design, need predictive metrics of how well a system's elements are expected to perform given a set of controlled conditions. That same program later in its life cycle (e.g., once fielding has occurred) might need a set of metrics focused more on providing information about overall system– and/or mission–level performance under a broad set of conditions. Other decisions, such as the adjustment of policies for compliance, need current and past performance metrics to reveal how well or poorly a previous policy worked. The full set of metrics needs to embrace both backward- and forward-looking views, with an emphasis that will, for a weapon system, vary depending on which phase of the life cycle it resides.

Ultimately, an absolute, quantitative assessment of survivability and effectiveness would be ideal. We do not see a way to do such assessments at this time. But we do propose a way to assess relative risk. Many decisionmakers face a need to compare alternatives. They might need to compare the performance of a program at different times. They might need to compare a system exposed to one threat environment versus another or compare the anticipated survivability and effectiveness of different designs for a system. Or, a senior leader might need to understand the expected performance of the enterprise before and after a proposed policy change. The comparisons should handle all forms of cyber concerns, including offensive attacks from outside, exfiltration of information, and insider threats. So, a framework generally needs to be able to adroitly support relative comparisons.

A final set of desired characteristics is that the metrics be part of a cohesive framework that helps indicate whether success in one measure partially offsets poor performance in another. In managing risk, decisionmakers will assess whether to take an action to mitigate risk or to accept risk based on these kinds of interdependencies.

All of these observations mean that a very wide set of metrics in an integrated structure is needed to fully understand how survivable and effective a weapon system might be in a cyber threat environment. Any given decisionmaker will need only a subset of these metrics.

Practical constraints prevent us from systematically achieving all of these desired objectives for metrics, but in developing the framework and metrics described in this report, these are the goals that we sought to fulfill.

Existing Approaches for Defining Metrics

The goal of this report is to define *metrics for weapon-system and mission survivability*[9] *and effectiveness in a contested cyber environment*. Rather than repeat this phrase throughout the report, we will often use the abbreviated phrase *cyber metrics*. For our purposes, a contested cyber environment includes offensive attacks by an outsider, exfiltration of information, and

[9] The Defense Acquisition University, 2015, defines *survivability* as "[t]he capability of a system or its crew to avoid or withstand a manmade hostile environment without suffering an abortive impairment of its ability to accomplish its designated mission."

insider threats within the cyber environment. All of these can, in different ways, put a system and the missions it supports at risk. The objective is to answer these questions: How can one estimate how survivable and effective a mission or weapon system might be in a specific cyber-threat environment given certain system design options, or policy options, or other comparisons? How can a program's cybersecurity and cyber resiliency be monitored over time? The goal is to establish metrics useful to specify system design requirements and to assess mission security and resiliency.

We start by discussing the measurement of survivability and effectiveness, then turn to the challenges posed by the cyber environment.

Specifying the survivability and effectiveness of a weapon system in a given threat environment is a ubiquitous task in the Joint Capabilities Integration and Development System (JCIDS) process. All JCIDS capability development documents, including the *JCIDS Manual*, contain a mandatory system survivability key performance parameter (KPP) "intended to ensure the system maintains its critical capabilities under applicable threat environments."[10] The KPP and other performance attributes, including key system attributes (KSAs)[11] and additional performance attributes (APAs),[12] are expressed by measures of effectiveness (MOEs)[13] and measures of performance (MOPs).[14] MOEs measure *mission* effectiveness and MOPs measure *system* performance.

Roughly speaking, MOEs state the mission objectives and MOPs are the parameters that can be adjusted to try to meet those objectives. Often the survivability and effectiveness can be formally described by a physical model. If the objective is for an aircraft to be able to release a munition in the proximity of an integrated air defense system, the output of the model might be how many aircraft on average survive and succeed in delivering munitions at some specified distance to the defense system (an MOE) given input parameters, such as the radar cross section of the aircraft and its speed (MOPs). The lines between MOEs and MOPs can blur, but the

---

[10] JCIDS, February 2015, including errata as of December 18, 2015, Appendix A, Enclosure D, p. D-A-2.

[11] KPPs are the most critical JCIDS performance requirements. Failure to meet a KPP might lead to program cancellation. KSAs are system performance requirements that are less critical than KPPs.

[12] APAs are the least critical JCIDS performance requirements. For an aircraft, typical APAs include weight and human-system integration requirements.

[13] The Department of Defense definition of *measure of effectiveness* is "[a]n indicator used to measure a current system state, with change indicated by comparing multiple observations over time" (see U.S. Department of Defense, July 2017). The Defense Acquisition University defines a *measure of effectiveness* to be:

> The data used to measure the military effect (mission accomplishment) that comes from using the system in its expected environment. That environment includes the system under test and all interrelated systems, that is, the planned or expected environment in terms of weapons, sensors, Command and Control (C2), and platforms, as appropriate, needed to accomplish an end-to-end mission in combat. (Defense Acquisition University, 2015)

[14] A *measure of performance* is a quantifiable measure of some system attribute meant to be testable. For an aircraft, speed, range, and payload are typical examples of MOPs.

overall sense is that MOEs are mission oriented and more general, and MOPs are more system oriented and more specific. MOPs should be precisely stated, preferably quantitative, parameters that can be put on contract and reasonably tested.

MOEs and MOPs are assigned two bounding values: threshold and objective. Performance below a threshold value either is not militarily useful or does not improve on existing capabilities. Performance above an objective value does not provide any additional military utility or is deemed not worth the additional cost.

Specifying the survivability and effectiveness of a weapon system in a distinct threat environment is a common task. A mandatory system survivability key performance parameter already exists in the JCIDS process. What, if anything, about this process is inadequate for developing metrics for the cyber threat environment?[15] In particular, what barriers are there to the definition of clear mission objectives (MOEs), parameters that govern a system's performance relative to those objectives (MOPs), and reproducible, validated, and verifiable models that assist in assigning desired values for these metrics?[16]

Peculiarities of Cybersecurity and Cyber Resiliency

Attributes

One of the slipperiest challenges of cybersecurity is that it is highly dynamic. Both the technologies and threat environment are constantly changing.[17] Not every aspect of the cyber realm is in flux, of course. Computer architectures, protocols, and some other large, structural elements have been much the same over decades. The fact of insider threats has long existed. However, key aspects of cybersecurity, such as peculiar technical vulnerabilities and their exploitation, depend on the details of implementation. Those details change, sometimes rapidly, over time. So even if, hypothetically, some exhaustive set of metrics were defined specifically enough to be characterized relative to a current threat, that set of metrics would likely prove to be dated well within the life span of a weapon system. On the other hand, metrics so general that they are timeless, such as generic access controls responding to known vulnerabilities, are

---

[15] Since 2014, a "cyber survivability endorsement" has been part of the system survivability KPP. Guidance for the implementation of the cyber survivability endorsement grapples with these issues and we use those insights throughout this report. At the time of writing in 2017, however, we are not aware of substantial program experience in using the cyber survivability endorsement in the Air Force.

[16] There is a small literature describing the challenges of measuring security. Many of these also propose some ideas for resolving some of these challenges, but none provide a comprehensive framework. For a summary of the challenges, see Roche and Watts, 1991; Anderson, 2008; Pfleeger and Cunningham, 2010; and Pfleeger, 2012. For some attempts at proposed frameworks and metrics that cover limited parts of the necessary scope, see Chapin and Akridge, 2005; Howard, Pincus, and Wing, 2005; Jansen, 2009; Pfleeger, 2009; Bau and Mitchell, 2011; Yee, 2013; Cheng et al., 2014; Wang et al., 2014; Hubbard and Seieren, 2016; and Zhang et al., 2016.

[17] See, for example, Kallberg and Cook, 2017.

unlikely to provide a complete picture of security given that the success of an attacker often hinges on the more dynamic details of a system, not on the overarching attributes.

A second attribute of the cyber environment is that it is complex. Many software systems are too complex for humans to understand thoroughly or to model or to test all possible permutations. For these complex systems, identifying and addressing all vulnerabilities is an unattainable goal. New vulnerabilities in software are discovered regularly; clever side-channel attacks against systems are frequently devised.[18] In fact, once software reaches a certain level of complexity, about the only thing one can be certain of is that some unknown vulnerabilities almost certainly exist. Complexity breeds flaws.[19] The consequence is that any set of metrics will have gaps due to ignorance. Also, the limitations of understanding complex systems mean that some degree of uncertainty in the assessment of metrics will persist.

A third attribute of the cyber threat environment is that it is ubiquitous in time and space. A weapon system is exposed to a cyber threat environment during wartime and peacetime, and while deployed and at home station (or even earlier, e.g., during design). There is no reprieve. Malware might be implanted in a weapon system (or supporting system) during peacetime at home base and activated later during wartime in a deployed location. Technical data exfiltrated during peacetime might be used during wartime to counter tactics. Any set of meaningful metrics must cover this large span of time and space. Unlike assessing the survivability and effectiveness of an aircraft near an integrated air defense system, where the threat is confined to that environment, assessing a cyber threat must cover the weapon system at home, deployed, during maintenance, and so on, all of the time. It is a much vaster scope than measuring the survivability against a kinetic attack.

A fourth and final attribute is the lack of a property. Cybersecurity and cyber resiliency have no firm underlying "laws of nature" comparable to kinetic threats.[20] For a typical measure of effectiveness not involving cyber, there is a physical basis that can be quantified, and the threat environment is generally more stable than the cyber threat. An aircraft can fly at a certain speed; a radar can track an object with specific characteristics; or a satellite can operate for a given duration in a specific radiation environment. These are physics-based attributes that can be modeled with confidence. They can be assigned quantitative values based on modeling and the expectation that the threat environment is more or less stable over the expected life span of the weapon systems. These specifications can be placed on contract with the vendor and tested during developmental testing. Because of a certain stability in the threat environment, operational testing in a noncyber threat environment is generally meaningful. Operational testing

---

[18] A side-channel attack is a method to exfiltrate information, usually encrypted, via indirect means. An example of a side-channel attack is to glean information being processed by monitoring the power used by a computer.

[19] Perrow, 1999; Shin et al., 2011.

[20] JASON, 2010.

is quite useful for the cyber environment but has limitations in being able to mimic all of the actions an attacker might take, now and in the future.

No equivalent to the radar range equation for modeling a radar system or laws of classical mechanics for modeling missile interception exists in the cyber realm. That cybersecurity and cyber resiliency, in general, lack this clean theoretical underpinning is related to the dynamic qualities, complexity, and ubiquity and emerges from these attributes. Securing a system or mission from cyber threats is inherently one of human-machine interactions and brings with it all the social complexities of human behavior of the attacker and the defender, as well as the technical complexities of modern software and hardware, all compounded by their mutual interactions.

Nevertheless, modeling can assist in assessing cybersecurity and cyber resiliency. Techniques and principles from a number of fields provide a rigorous basis for assessing some questions in some specific, albeit sometimes limited, domains.[21] Some examples include model checking of code via formal methods, identification of critical nodes and links using centrality metrics in network analysis, methods in cryptanalysis for weaknesses in encryption, and so on. Qualitative methods also can contribute, including studies of individual and organizational behavioral analysis from sociology and psychology. We leverage these in this report, but it is necessary to understand their strengths and their limitations, and that each of these addresses a very limited scope of what needs to be known.

Implications

The enterprise objectives, needs of decisionmakers, and attributes of the cyber environment are not just challenges. They are also guides for constructing a framework for cyber metrics and defining those metrics.

One of the key attributes of cybersecurity and cyber resiliency is that the technology and threat change rapidly relative to the life cycle of a weapon system. This attribute, in combination with the complexity of the technology and the human-machine interaction, demands that solutions to survivability and effectiveness in a contested cyber environment be responsive. No forethought, no matter how carefully done, will suffice in anticipating all the possible moves an adversary might take in the future. Measures of effectiveness, then, need to reflect the ability of the hardware, software, people, and processes to be adaptive to the threat environment as it evolves. Metrics need to reflect how well the U.S. Air Force can play a two-player cyber game.[22] We call these the *dynamic* dimensions of cyber metrics.

There are, nevertheless, domains in cybersecurity and cyber resiliency that do not change in nature rapidly (e.g., protocols). For these, some sound policies might be known, such as security controls for limiting unauthorized access. While security controls are in general insufficient, they

---

[21] JASON, 2010. For some proposed examples of quantitative measures, see Shi et al., 2016, and Zhang et al., 2016.

[22] Alpcan and Başar, 2011.

are good practice. Known good practices should also be captured. We call these *static* dimensions of cyber metrics.

Because of the complexity of the problem, the ubiquity of the threat, and the lack of underlying "laws of physics" governing the cyber environment, performance in no single domain will completely express the soundness of cybersecurity and cyber resiliency. Metrics need to span home base and deployed, peacetime and wartime, all ware (soft, firm, and hard), people, organizational performance, and processes. A complete picture, or as complete as can be obtained, will emerge only from a combination of many metrics across many areas. Most decisionmakers will need a portfolio of related metrics and a way to understand how they work together.

Also because of the complexity of the problem, the ubiquity of the threat, and the lack of underlying "laws of nature" governing the cyber environment, few of the metrics themselves are likely to be perfectly well defined, in the sense that they measure exactly what the decisionmaker thinks or wants them to measure. This limitation indicates a need for metrics, or feedback in general, on how well the metrics themselves are performing so that they can be adjusted and uncertainty is better understood. The assessment of cybersecurity and cyber resiliency needs itself to be assessed.

Shaping a Framework for Cyber Metrics

For cyber survivability, there are two overarching activities to be measured. The first largely comes from the desire for cybersecurity and is negative in nature: preventing the adversary from doing something. The second largely comes from the desire for cyber resiliency and is positive in nature: ensuring that missions continue even when under cyber assault. With those activities in mind, the needs of the decisionmakers and the peculiarities of the cyber environment combine to suggest the outlines of a framework for cyber metrics.

Adversaries are continually developing new strategies and tools to gain some advantage in the cyber domain. Defending against these onslaughts requires an equally clever set of counterstrategies and tools. This situation can be profitably viewed as a two-player game. (Because the word *defense* in this context elicits thoughts that exclude resiliency and adaptive behaviors, henceforth we will not refer to the actors in the two-player game as an attacker and a defender but by the more generic terms *Red* and *Blue*, respectively). A framing question to identify and associate metrics into a cohesive structure is: What Blue actions would disrupt Red's ability to attack a weapon system through cyberspace and negatively affect a mission? This view emphasizes the moves of Red and adaptive responses to those moves by Blue, and contrasts with frameworks that take a defensive view focusing only on actions of Blue.

The most common frameworks for cyber metrics do not highlight Red-Blue interactions. They emphasize, for example, the need to preserve confidentiality, integrity, and availability of data,[23] or the need to identify key assets, protect them, detect intrusions, respond to those intrusions, and recover from any ill effects.[24] In the cat-and-mouse game of cyber operations, this view neglects the actions of Red, thereby limiting the scope of assessment. As we argue in the next chapter, cyber metrics based on the Red-Blue interplay expand this scope by providing better insight into whether actions taken by Blue are, or are not, sufficient, and whether poor performance in one measure might be partially offset by good performance in another.

The two-player game paradigm implies two tiers of metrics: one of actions that Red needs to take to be successful, and ones that Blue needs to take to thwart Red. The former are mission objectives and have many of the characteristics of MOEs. The latter are more detailed actions that Blue can take as mitigations to Red's actions and have many characteristics of MOPs. Some of these Blue actions will be preplanned and deliberate (cybersecurity actions), but many will be reactive to Red moves, and thus be dynamic and adaptive (cyber resiliency actions). The latter set of dynamic and adaptive Blue actions must be measured by how much they enable the system and its operators to adjust to a changing cyber threat environment and recover from any attacks.

This view is less pessimistic than some statements about cybersecurity that assert that the offender only needs to find one avenue of attack and that the defender needs to succeed at finding and addressing all vulnerabilities. In the two-player game paradigm, Red must coordinate and execute a number of activities to be successful (have a mission impact on Blue that is worth the resources that Red must expend). Blue just needs to interrupt enough of these activities or make them sufficiently costly that Red is not successful.

Blue's task is still difficult, but not insurmountable. Quite a range of Blue mitigations will be needed that span all dimensions of the threat, including home base and deployed, peacetime and wartime, all ware (soft, firm, and hard), people, organizational performance, and processes. These mitigations should be, as best as possible, aligned with existing boundaries of authorities of the decisionmakers they are to support. Most of these metrics will be at the working level to support decisionmakers at the working level of the organization, such as engineers. We develop this aspect of the framework in detail in Chapter 2.

On top of these metrics is a need to assess the metrics themselves and how well the institution supports the countermeasures that they assess. How well are the metrics themselves capturing past and potential future performance? Could excellent performance in the areas monitored still lead to Red prevailing over Blue? What needs to be monitored to assess these assessments, and how can that be done? We take up these more complex meta topics in detail in Chapter 3.

---

[23] See, for example, Stoneburner, 2001.

[24] See, for example, U.S. Department of Homeland Security and Executive Office of the President of the United States, 2015.

2. Monitoring at the Working Level

> "One can sympathize with the general who shouts at his analyst: 'Give me a number, not a range': Unique numbers are easier to work with and to think about than ranges or probability distributions. But he is probably asking his analyst to falsify the real world in a manner that will make it impossible for him, as a commander, to make good decisions."
>
> — Charles J. Hitch and Roland N. McKean[1]

This chapter presents a range of working-level metrics of the survivability and effectiveness of a system or mission in a contested cyber environment. They are meant to be, to the level of indenture given, exhaustive in terms of Red actions to be foiled. We also present a number of potential Blue countermeasures but emphasize that this list of Blue actions can never be complete. The overall tasks of cybersecurity and cyber resiliency are not fixed but are dynamic. Therefore, Blue countermeasures must be vigilantly and constantly reassessed given environmental factors, such as Red's posture and technological changes.

All cyber metrics are presented in a framework. The framework is an important construct for two reasons. First, it helps ensure that the Red actions to be addressed are comprehensive. Second, it places all metrics in a context that, at least to a limited degree, helps identify whether success in one area might partially offset poor performance in another. A framework of this nature is necessary to eventually make wise resource allocation decisions.

We do not, in this chapter, address the possibility that any gaps in these metrics might exist (e.g., in the institutional seams between programs). These and other institutional-level factors that might impede Blue's mission effectiveness are treated in detail in the next chapter.

Defining a Cyberattack Path

The survivability and effectiveness of a mission or a weapon system in a threat environment can be viewed from both offensive and defensive perspectives: What does Red want to attack, and how does it need to attack to be successful? What does Blue need to do to thwart these attacks? Despite being complementary views of the same problem, the different perspectives can yield distinct analytical insights.

Let us consider these two perspectives for cybersecurity. From Blue's viewpoint, we might list all the actions believed to help defend and protect the weapon system. Examples would include means of detecting intrusions, searching for and mitigating vulnerabilities, and identifying ways for a system's functions to recover after an attack. For each action, some

[1] Hitch and McKean, 1960, p. 188.

measure of merit would be developed against which the current state of the system would be assessed. But, how good does each action need to be? How does the effectiveness of one action bear upon the need for a certain level of effectiveness of another action? The breadth of the goal—to defend against attack and persist in operations while under attack—does not help much to reveal these interdependencies. We are left with trying to score as high as possible on the scales of all metrics with little insight into how good each needs to be. Also, how do we know whether we have measured everything needed to assess success or failure?

To achieve these goals, the framework needs to take a mission perspective and be as comprehensive as possible. Other frameworks for modeling Red's attack path through cyberspace have been offered. Most are developed to describe how Red must operate at a technical level. The framework we outline is broader, encompassing everything that Red must do to be successful against a mission, not just the technical aspects.

Looking at the issue from the Red's perspective, we would start with what Red's objectives might be in attacking Blue and then list all of the actions in the attack path that Red would need to execute to be successful. By successful, we mean the ability to interfere with Blue's mission execution to an extent unacceptable by the Blue. Examples might be to gain access to the system, gain enough knowledge of the cyber system to create a desired effect, muster the resources to pull off the attack, and use the cyber effect to create some negative mission impact. Again, for each of these actions, some measure of merit would be developed against which the current state of the system would be assessed. The advantage of this offensive perspective is that, by being clearer in the objective—disrupting the Red's attack path—this perspective gives a deeper insight into how good each action needs to be.

To illustrate this insight, we look at a simple example of part of the attack path. At the highest level, Red must do four things: (1) access the system in question (access) *and* (2) know enough about it to execute an attack (knowledge) *and* (3) have the resources and capability to carry out the attack (capability) *and* (4) create an effect that has significant negative mission repercussions to the defender (impact). More specifically, by *impact* we mean that the cyber effect imparts real, negative consequences to the targeted mission. These four actions do not need to be sequential. They might occur in a different order for different attacks and might overlap in time. The first three actions (access, knowledge, and capability) can be thought of as costs that Red must overcome and the fourth (impact) as the degree of benefit accrued to Red.

Now, seeing the problem from Red's perspective, we can see how effort expended by Red is distributed and how a set of mitigation efforts might or might not combine to upset a Red cyber operation. Suppose the impact that Red desires to achieve is to defeat a certain critical mission element of Blue—say, provide aerial refueling for theater operations. For simplicity, further suppose that Red has two plausible ways to defeat these operations by a cyberattack: disable the refueling boom on the deployed refuelers in the theater or disrupt the fuel operations at the bases from where the refuelers are to be launched. Assume further that from Red's perspective these modes of attack deliver roughly the same impact, and that Red has both the needed access and knowledge to conduct either. However, let us say that one of them (attacking the booms) is considerably more challenging and requires far more resources.

Because Red's cost-benefit ratio in this example is more favorable for attacking the fueling capability on the ground rather than the boom on the aircraft, it is more likely that Red would choose to attack the fueling capability on the ground. It is of more value then, in this example, for Blue to have sound mitigations for the attack to the fueling capability on the ground than the attack on the booms, all other things being equal. The more resources that Red needs to expend for a given effect, the less likely Red will select that method.

There is another way in which the offensive view provides useful insight—the attack path provides a relational framework showing how the strength of one mitigation effort by Blue might offset a weakness in another mitigation effort. This framework emerges from the fact that some actions that Red must take must all be done, and, in other cases, Red has options. The four high-level actions shown in Figure 2.1 are linked by Boolean *and* statements because each must happen or the attack is thwarted. Because Red must do all four to some degree, reducing the likelihood of success of one or more of them increases the likelihood of interrupting the attack path. The greater the number of these four actions that the defender can inhibit, and the greater the confidence in inhibiting each, the more difficult the job of the attacker and the more survivable and effective the system is in that cyber environment.

Figure 2.1. Red's Cyberattack Path Decomposed for Access and Knowledge

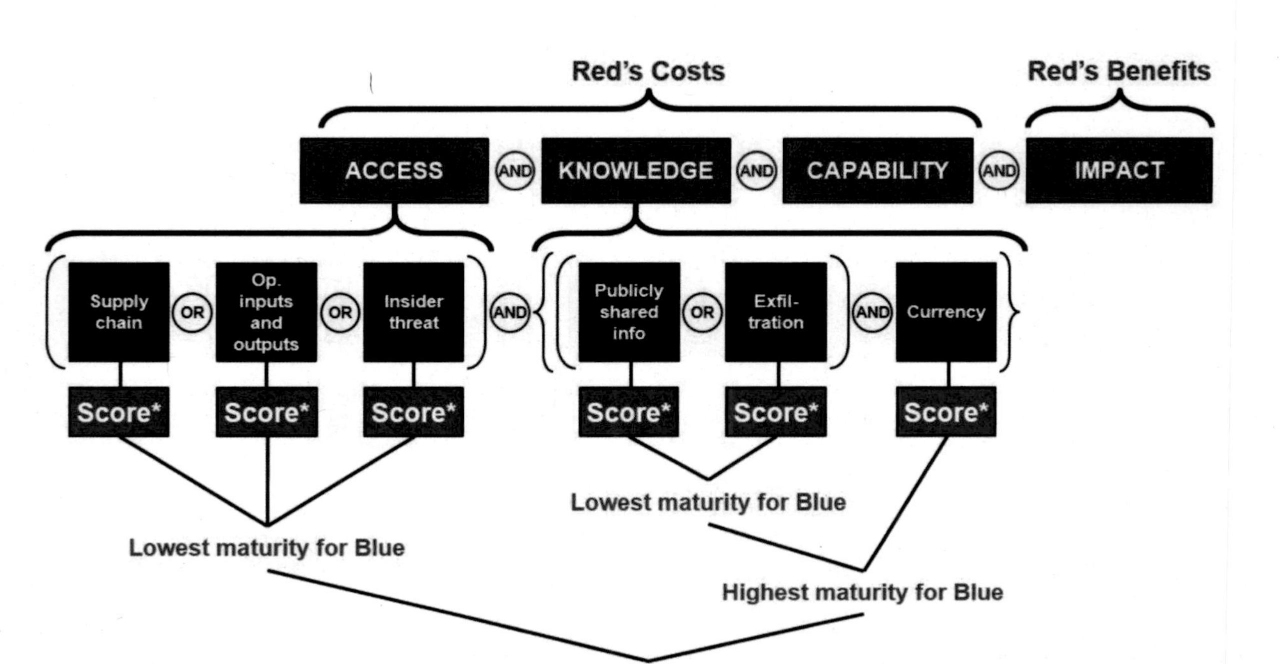

Each of the four actions (access, knowledge, capability, and impact) can be further broken down into subordinate activities. For example, access might be gained through the supply chain, through the inputs and outputs of data of a weapon system, or through a malign inside actor. Sufficient access to the system might be gained from any of these three groups of approaches, and hence they are connected by Boolean *or* statements. Each of these activities can, in turn, be further divided. For example, access to the supply chain can be garnered through design, manufacturing, transportation/distribution/installation, maintenance, and disposal. These are also connected by Boolean *or* statements because access, in general, needs to be gained in only one of these phases.[2]

The power of the attack-path approach lies in the richness of the Boolean structure of the breakdown of the attack path into finer elements. To illustrate this approach and how it reveals the relative value of scores of differing metrics, let's look at a partial breakdown of the access and knowledge parts of a cyberattack path, depicted in Figure 2.1. Access, as mentioned previously, can be broken down into supply chain, inputs and outputs of data of the weapon system, and insiders. Knowledge can be broken down into two domains of knowledge. The first is information that is publicly available (such as commercial, off-the-shelf technologies) and publicly released information (e.g., contracting information). The second is information that is held secret and must be exfiltrated, such as government-unique software. However, there is a

---

[2] Red might prefer to access by one of these mechanisms over another. We take up this point later in this chapter.

third constraint for Red to overcome—the knowledge gained has to be fresh enough to be relevant for the attack. We call this need *currency*.

The usefulness of knowledge for an attack depends on its currency. Red needs to act on knowledge it has collected before Blue makes any relevant changes to the targeted weapon system. If, for example, Red needs to know the specific configuration of a router or the location of a weapon system, that information has to remain stable between the last time Red collected it and the time Red uses it for an attack. Blue needs to stay inside Red's observe, orient, decide, and act (OODA) loop.[3] Blue can stay inside of Red's OODA loop by actively responding to Red's actions and changing system configurations accordingly, or fortuitously by deliberate resetting of systems so as to increase the burden on Red to maintain adequate current knowledge.

Putting these actions together, Red needs specific knowledge, gained either through publicly released information *or* by exfiltrating the information *and* the information must be current, i.e., the targeted system must be sufficiently stable for the attack to be successful. This level of decomposition of knowledge, therefore, presents a combination of both Boolean *and* and *or* connections.

Applying the Boolean logic to this structure shows how assessments of the ability to inhibit each of these activities can be mutually compared. Leaving aside for the moment how to assess the efficacy of each of these actions (a topic we take up shortly), assume that we have some score for each that allows a comparison of scores across these different concepts. For any group that is connected by a Boolean *or* connection, such as two of the elements of knowledge (publicly known or exfiltrated), we would assess this group as being the *worst* (for Blue) of the pair. For any group that is connected by a Boolean *and* connection, such as the knowledge gained and currency of that knowledge, we would assess the group as the *best* (for Blue) of this pair. This logical roll-up extends through the whole of the assessment, as depicted in Figure 2.1. The roll-up applies to as fine a decomposition of actions as needed. For the case of the supply chain, it would be applied to the level of design, manufacturing, transportation/distribution/ installation, maintenance, and disposal avenues for supply-chain access.

What to Measure

Red Actions

As argued earlier and diagramed in Figure 2.1, the principal actions that Red needs to achieve in cyber operations are to have access to a target, to have enough knowledge of the target to succeed, to have the capability (resources) to carry out the operations, and to have whatever actions Red achieves create meaningful impact on Blue's missions. We discuss these four actions

[3] The OODA loop idea was coined by John Boyd. The central idea is that the player, Red or Blue, that can observe, orient, decide, and act faster than the other can, gains the initiative and keeps the adversary confused. Boyd never formally published his work; for a discussion, see Freedman, 2013, pp. 196–201.

in this order, breaking down each into higher levels of indenture from Red's perspective. The categorization of the Red actions is meant to be, as best as we can do, exhaustive. In making any list like this one, there is a risk of overlooking an element. If the reader sees the need for an additional category of Red actions, it can be added to this list. The categories are not absolutely sharp, so it is possible that a specific Red action might be placed in more than one category. The point of this list is to establish the goals of Blue actions. By identifying Red actions that potentially need to be countered, and by assessing how well each of the actions are addressed by Blue countermeasures, insight is gained into the overall cybersecurity and cyber resiliency of a mission or system.

In the following section on Blue countermeasures, we highlight ways in which Blue might respond.

Access

There are essentially three main vectors for gaining access to a system: through the supply chain, through a system's pathways for data inputs and outputs, and through an inside actor.

Supply Chain

Entry points of Red into the supply chain are during design, manufacturing, shipment and storage, maintenance, and disposal. Design is especially important in the cyber domain. Many vulnerabilities are due to buggy code.[4] Maintenance is also a critical phase because of the plethora of connections a system has, sometimes intermittently, to various maintenance equipment. Vendors own and control some of this equipment. Also, maintenance persists throughout a weapon system's life cycle, making it a continuous target at all times and locations. Even disposal poses a risk: Improper sanitization of components during disposal can leak critical information about systems and their architecture to an adversary.

Data Pathways

Data pathways are the designed ways in which a system exchanges data externally. These include direct communications connections, noncommunications data apertures (antennas, sensors), air-gapped media, and connections to subsystems. The latter includes side-channel attacks, electronic emanations (specifically, TEMPEST), and other indirect access vectors.

Insiders

Insiders are human agents who can grant access to Red. They might be inadvertent Blue actors (those who make an unintentional mistake that grants Red access) or malicious actors (Blue moles or Red spies). For some countermeasure techniques, it is useful to distinguish Blue moles as either disgruntled, recruited, or self-motivated compromised actors.

---

[4] Stolfo, Bellovin, and Evans, 2011, p. 63.

This category overlaps with several of the other Red actions. For example, the insider might act in the supply chain, or help gain knowledge about a system or mission. We list this category separately, however, because some Blue countermeasures will need to be crafted specifically to the insider threat and it might be overlooked as part of responding to other Red vectors.

Knowledge

Knowledge is the information that Red needs to *effectively* carry out cyber operations. That information can be in the public domain or be classified. A further, very important dimension to Red's knowledge is that, to be effective, the information must also be sufficiently accurate and current.

Public Information

Information that is essentially public includes commercial, off-the-shelf ware, information gleaned from open source intelligence (OSINT), and information offered in public requests for proposals and contracts.

Protected Information

Red can exfiltrate information that is meant to be secret through signals intelligence, which includes communications intelligence (COMINT), electronic intelligence (ELINT), and foreign instrumentation signals intelligence (FISINT); through human intelligence (HUMINT); and from information extracted from bought or captured ware (foreign military exploitation).

Currency

Red needs the information that it gains to be accurate and current. Currency is compromised if it takes Red too long to collect, disseminate, and analyze the information (and conduct the cyber operations) relative to the timescale of changes to the Blue target. Red needs to stay inside Blue's OODA loop.

Capabilities

Red must have the capabilities to perform the cyber operations. The quantities and qualities of capabilities that Red requires to be successful depend, of course, on Blue. Blue can design and operate in ways that make Red's task hard, easy, or somewhere in between. For simplicity, we list in this category just those capabilities that Red controls. We include Blue-controlled attributes that can elevate the capabilities that Red requires in the next category (impact).

Infrastructure

Red needs infrastructure to conduct cyber operations. The type of infrastructure depends on the cyber operations method and the target. One example is the use of a botnet to conduct a distributed denial of service attack.

"Orange" Actors

Red can also benefit from the assistance of other parties sympathetic to Red's objectives, often called "Orange" forces in wargames. Orange forces include alliances with other nation states and surrogates (such as criminal elements acting on behalf of Red), and tend to have access to markets for procuring malware or cyber-operations services that Red either does not have the capabilities to do organically or prefers to outsource to mask attribution.

Manpower, Personnel, and Training

Red needs the workforce, the skilled and loyal people (not susceptible to their own insider threats), and training to succeed in cyber operations.

Financial Resources

All operations require money, and Red needs sufficient funding to succeed with cyber operations.

Impact

Blue can make Red's task inherently difficult by raising the bar of capabilities that Red must muster to succeed in its cyber operations. Red's impact assessment will include such factors as the importance of the mission (the priority from Red's perspective), the assessed degree of mission impairment from the attack, the assessed duration of impairment, the likelihood of being detected (i.e., avoiding attribution), the probability of success, and whether any collateral damage that might occur is unacceptable to Red. In some circumstances, Blue can influence one or more of these factors.

Resiliency is the main way in which the operational impact of Red cyber operations is reduced by Blue. Cyber resilience issues from architectural choices at both the mission and system levels. At the mission level, a robustly designed mission architecture, verified and validated through cyber mission thread analysis, is the principal means for achieving cyber resiliency. The emphasis at this level is on mission concepts of operation and interactions among systems.

At the system level, cyber resiliency arises out of a number of attributes. One attribute is to stipulate in designs that any safety- or mission-critical function be provided by more than one subsystem such that no single cyber vector of attack could simultaneously degrade the functionality of all of the subsystems that provide that safety or mission-critical function, a property we call *cyber separability*. Another attribute is to design computer code in which it is, to some extent, mathematically provable that the code cannot perform functions other than those it is required to do, achieved for code of permissible size via *formal methods*.[5]

---

[5] See, for example, Warwick, 2015; Anonymous, 2015; and Degabriele, Paterson, and Watson, 2011.

Blue Countermeasures

The effects that Red aims to inflict are, of course, not inexorable. Blue can try to thwart Red using a range of countermeasures. No fixed and comprehensive set of Blue countermeasures exists, or will ever exist, to ineluctably thwart Red. True to the cat-and-mouse game being played, Red will develop new capabilities and its tactics will evolve as new concepts of operations and technologies emerge. In response, Blue will need to continually re-evaluate countermeasures and perhaps devise new ones. The categories of Red actions to be curtailed will be relatively stable; the Blue countermeasures will be less so as they adapt to evolving technologies and the dynamic threat environment.

Blue leadership must remain alert to limiting their actions to following lists of prescribed countermeasures. Such lists, of which the security controls in the National Institute of Standards and Technology (NIST) Risk Management Framework are a prime example, are meant to be guides to assist thought.[6] The goal of lists of potential countermeasures is to aid Blue in finding the right combination of actions. Lists of countermeasures are not substitutes for thinking. Blue personnel, rather than avoid the distraction of mere compliance with lists of candidate countermeasures, must keep their eyes on the goal—thwarting Red—and the countermeasures employed must be evaluated according to how well they are doing in thwarting Red.

With these caveats in mind, we note some candidate Blue countermeasures. Effective mitigation for any Red action is most likely to arise from multiple Blue countermeasures and by multiple actors (e.g., program office, operator, and so on). Our goal in this section is not in any way to be comprehensive. We draw attention to selected Blue countermeasures to make some specific points.

Countering Access

Countering the cyber supply chain threat is well discussed.[7] Common methods include component inspection; positive inventory control; directed actions in maintenance technical orders; trusted shipping and warehousing; antitamper techniques; trusted foundries and other suppliers; and procedures for sanitization and disposal of articles at their end of life. Many of these measures are arduous and must be reserved for the most critical components. Some also depend on an industrial base loyal to the United States. Preserving such an industrial base is a government-wide responsibility.

Countering access via standard data pathways is the focus of most cybersecurity efforts in the U.S. government.[8] We add only a few observations often not treated in cybersecurity guides. Blue must be ever vigilant in establishing countermeasures to side-channel attacks. Even

---

[6] National Institute of Standards and Technology, 2015; Committee on National Security Systems, 2014.

[7] Boyens et al., 2015; Shackleford, 2015.

[8] Much of the work by the National Institute of Standards and Technology, 2015, and Scarfone and Mell, 2007, offers methods for countering this Red vector.

provably secure measures, such as RSA encryption algorithms, can be broken by side-channel attacks.[9] Side-channel attacks lurk in some surprising places.[10] In the specific case of access through radio-frequency links and some sensors, electronic warfare techniques can be useful. Techniques, such as application whitelisting, are useful both for countering access through data pathways and for a range of other Red actions, including some supply chain access vectors.[11] Although widely used for security, encryption offers an additional layer of defense against access, but its reach is not universal, especially for data at rest.

Countering the insider threat illustrates a common attribute of Blue countermeasures—they are often only partially successful at countering a particular Red action, but they can be quite useful for a range of potential Red actions. Common countermeasures include limiting access to least privilege, segregating networks, personnel screening, training, deterrence, continuous monitoring of networks and personnel, and maintaining a positive work climate.[12] None of these countermeasures are sufficient in isolation. Yet in combination, they form a powerful lattice of actions to prevent and limit the effects of an insider. They are also good practices for countering a range of other Red actions, such as insider threats through the supply chain.

Countering Knowledge

The main countermeasure for limiting Red's knowledge gained through "publicly" available information is by carefully reviewing what information should be released. One salient example is in limiting the information released in requests for proposals. For example, there is often no reason why the contractor bidding needs to know Blue's exact use and location of some equipment. Revealing this kind of information unnecessarily expands the knowledge base of Red.

Countering COMINT, ELINT, FISINT, and HUMINT largely falls under counterintelligence, which lies outside of the focus of this report. The acquisition and test communities do, however, have a role in limiting the knowledge gained by Red through FISINT. Countermeasures include operational security measures and encryption. To some degree, the acquisition community also plays a role in all of these collection areas by determining at what level information is to be classified and handled.

The accuracy of the information available to Red can be compromised by Blue using deception and employing decoys. Deception and decoys, such as honey pots, apply to both publicly available knowledge and controlled knowledge. These countermeasures contribute to confusing Red, and thereby diminish the confidence in and currency of Red's knowledge.

---

[9] Degabriele, Paterson, and Watson, 2011.

[10] One recently discovered example is that design information can be stolen from the sounds emanating from additive manufacturing tooling. For more information, see Hvistendahl, 2016.

[11] Sedgewick, Souppaya, and Scarfone, 2015.

[12] Cappelli, Moore, and Trzeciak, 2012; CERT Insider Threat Center, 2016; Bunn and Sagan, 2016.

Changes to Blue systems or missions that erode the currency of Red's knowledge is an interesting countermeasure to Red's knowledge and is insufficiently discussed in cybersecurity and cyber resiliency policies and practices. Blue can nullify the usefulness of Red's knowledge by making it obsolete. There are various ways to diminish the stability of Blue's posture, including recapitalizing and modernizing systems, changing concepts of operations, using war reserve modes, rekeying encryption, and sometimes simply rebooting an operating system.

Countering Capability

Most of the techniques for countering Red's capability lie outside the responsibilities of the acquisition community. Nevertheless, they are another weapon in Blue's countermeasure quiver that can be most effective in some circumstances. Some of these countermeasures include active defense (e.g., to interrupt Red's distributed denial of service infrastructure); alliances; legal frameworks; international norms; deterrence (e.g., by sharing infrastructure with Red such that the collateral damage is unacceptable to Red); and by both kinetic and nonkinetic attack of Red's resources (e.g., infrastructure, people).

Cyber Metrics

The culmination of the arguments that we have made is that to measure how survivable and effective a mission or system is in a contested cyber environment, we must understand how well Red cyber operations are being countered. The focus of cyber metrics, then, must be on Red's estimated success or failure, not on the specific countermeasures that Blue might try. Blue countermeasures are important, of course, but their importance is as a means to an end—that of hindering or thwarting Red. Cyber metrics that concentrate on the compliance with lists of candidate Blue countermeasures fail to indicate whether those measures are effective, properly implemented, or sufficiently comprehensive to thwart Red. Therefore, the compliance approach is insufficient.

What to measure, then, is how well Blue is countering the Red vectors listed earlier. How well are Red actions during the design phase of the supply chain being met? How well is the potential Red access to a system via subsystems being countered (e.g., maintenance test equipment)? How well is the insider threat being addressed? On what basis are the decisions about the guarding of information (e.g., publicly releasable, controlled unclassified, and at the various classification levels) being made? How stable are aspects of the system or mission that Red needs to know to successfully attack relative to Red's intelligence refresh cycle? The answers to these questions, addressing the full range of Red actions, form the set of working-level cyber metrics. These cyber metrics require knowledge and expertise of a skilled workforce and are not easily reduced to automated collection.

How to Measure

As argued in Chapter 1, cyber metrics are, in general, more complicated than measuring the effectiveness of countering kinetic threats. Cyber metrics cannot generally be reduced to quantitative models like those used to assess the effectiveness of, say, maneuverability, speed, or low observability in penetrating an air defense system. They must, for example, also include how humans interact with ware, and whether Blue's responses to Red are sufficiently timely. Any assessments of the effectiveness of Blue countermeasures to Red cyber operations will be inherently uncertain and require vigilant, continuous attention by skilled workers.

The focus on thwarting Red, rather than compliance to scripted Blue countermeasures, also means that assessments must be relative to the baseline of the Red threat. If Blue's cybersecurity and cyber-resiliency countermeasures are advancing, but Red is advancing faster in its capabilities and tradecraft, the improvement in Blue cyber metrics is illusory. Blue has actually lost ground. This characteristic is not unique to cybersecurity and cyber resiliency. It is true of all assessments of military power.[13]

What we propose, then, is a three-tiered assessment: (1) an assessment by responsible leaders for a weapon system and supported missions of how well each Red action is countered; (2) some external assessment (a verification and validation) of each of these self-assessments; and (3) a missionwide roll-up of all of the countermeasures, looking for how effective they are as an ensemble against each of Red's actions, and whether any gaps exist.

Assessing at the Working Level

The locus of information about the technical details of weapon systems and missions lies at the working level. As argued earlier, cybersecurity and cyber resiliency are technical and highly nuanced. Some of the ideas for effective Blue countermeasures will only be discovered at the working level. Many deficiencies will only be evident at the working level because the relevant knowledge exists only at that level. Also, the nimbleness to respond to Red's evolving tactics is facilitated by empowering working-level actors. Self-assessments at the working level are critical to identifying the right countermeasures and assessing how well they are expected to hinder or thwart Red.

A decisionmaker needs two kinds of information to value a decision regarding how well Blue is doing: the probability that the decision will lead to a certain outcome, and the value of that outcome.[14] In our context of working-level cyber metrics, this means determining both the expected effect that a Blue countermeasure will have on a Red action and the uncertainty in that assessment.

---

[13] See, for example, Marshall, 1966.

[14] Churchman and Ackoff, 1954.

The expected effect of Blue countermeasures on Red cannot, in general, be assessed quantitatively. What is possible is to make a coherent argument for which countermeasures should be employed and what their expected efficacy might be. There is an analogy here with verification and validation of modeling and simulation software. The burden of proof for software verification and validation lies with the software developer, who documents all of the various tests that the software has been subjected to in order to demonstrate that it most likely performs as it should.[15] For assessing working-level cyber metrics, each key leader (program office, operational wing, etc.) at the working level would similarly document all the steps that he or she has taken to counter each of Red's actions. Because each leader's remit is limited, there will be Red actions for which no countermeasures have been taken, and some for which the countermeasures might be weak (with the expectation that others will compensate with additional countermeasures).

Because such assessments are inherently subjective, the assessment valuation must reflect this uncertainty. The best that can be done is some kind of range, qualitatively specified, for most metrics. We recommend assessing the ability to address Red actions by a maturity index.[16] An immature response to a Red action would be limited to static dimensions of Blue countermeasures. Static (cybersecurity) actions would progress in maturity if the following questions are addressed in the affirmative:

- Have static countermeasures been identified?
- Have static countermeasures been implemented?
- Have static countermeasures been tested?
- Have static countermeasures been found to work adequately?

For example, in countering Red access to standard data pathways to a system, has the program office identified all the pathways? This identification is more difficult than it might seem, as pathways might exist in commercial, off-the-shelf components that a contractor uses, but for which all of the technical details are not available because of how the contract was written. Once all pathways have been identified, have mitigations been identified? If so, have they been implemented? If implemented, have they been tested (e.g., by a certified red team)? The deeper the answers are affirmative in this sequence, the more mature the countermeasures.

A yet more mature response would include dynamic (cyber resiliency) actions, including measures to decrease the impact of Red's cyber operations (e.g., robust operational architecture, cyber resiliency, use of formal methods). The deeper the following questions are assessed in the affirmative, the more mature:

- Has a baseline trusted state been defined for each system to return to?
- Is continuous monitoring done?

---

[15] Department of Defense Instruction 5000.61, 2009.

[16] Modeled after U.S. Department of Energy and U.S. Department of Homeland Security, 2014.

- Have resilient methods been identified to return to this state?
- Can these actions respond within Red's OODA loop?
- Is the recovery time adequate for mission needs?
- Have these measures been implemented?
- Have they been exercised and tested?
- Have they been found to be adequate?

All of these assessments would be gauged against Red's capabilities. Given the ubiquity of the cyber threat, this would be the highest Red actor assessed by intelligence. Assessing the ability to address each of Red's actions by a maturity index defined in this way places all assessments on a common scale and does not push the scoring to an unjustified quantitative precision.

The most important product of the assessment at the working level would be artifacts that document what is being done to counter each of the Red actions. It is at this level where the technical and operational details lie. This level must do the most detailed assessment to document the Blue countermeasures and their qualities against Red. The leadership to which the working level reports will, however, need to see some aggregation in the form of a scoring of these assessments. This aggregation is necessary to get a broader view of the mission without needing to review all of the artifacts produced at the working level. We recommend a maturity index for this scoring, as described in Table 2.1.

Scoring of the maturity index for each Red vector should be done conservatively. To progress upward to higher maturity scores, issues at each maturity level must be addressed as completely as feasible. One weakness in Blue's countermeasures against a Red vector can be all that Red needs to prevail. Therefore, even if some countermeasures against a Red vector have been identified, implemented, tested, and found adequate, if others have not been properly identified, the response to this Red vector remains immature. For example, if countermeasures for accessing a weapon system through communications paths have been comprehensively identified, implemented, tested, and found adequate, yet all the paths through which Red might gain access via subsystems remain to be identified, the scoring for Red access via data pathways remains "most immature."

Table 2.1. Maturity Levels for Working-Level Metrics

Level	Maturity	Characteristics
1	Highest maturity	Solutions to counter a Red vector are tested, exercised, and found to be adequate.
2	Mature	Solutions to counter a Red vector are implemented with continuous monitoring.
3	Intermediate	Solutions to counter a Red vector are identified.
4	Immature	How Red might act by a vector is understood in the context of the system or mission under review, and a baseline trusted state is defined.

Level	Maturity	Characteristics
5	Most immature	Awareness of how Red might act by a vector against a system or mission is inadequate. Such examples for access include incomplete knowledge of standard pathways for data inputs and outputs of a system.

Shown graphically, the end result of this assessment would look something like the diagram in Figure 2.2. The figure shows a notional assessment for six selected Red actions by one working-level assessor. Assessments for each Red action have two components: a static (cybersecurity) assessment, shown in blue, and a dynamic (cyber resiliency) assessment, shown in green. Each of these consists of an aggregate of all of the Blue countermeasures under the control of the assessor.

Figure 2.2. Notional Example of Maturity Indexes for Countering Selected Red Actions

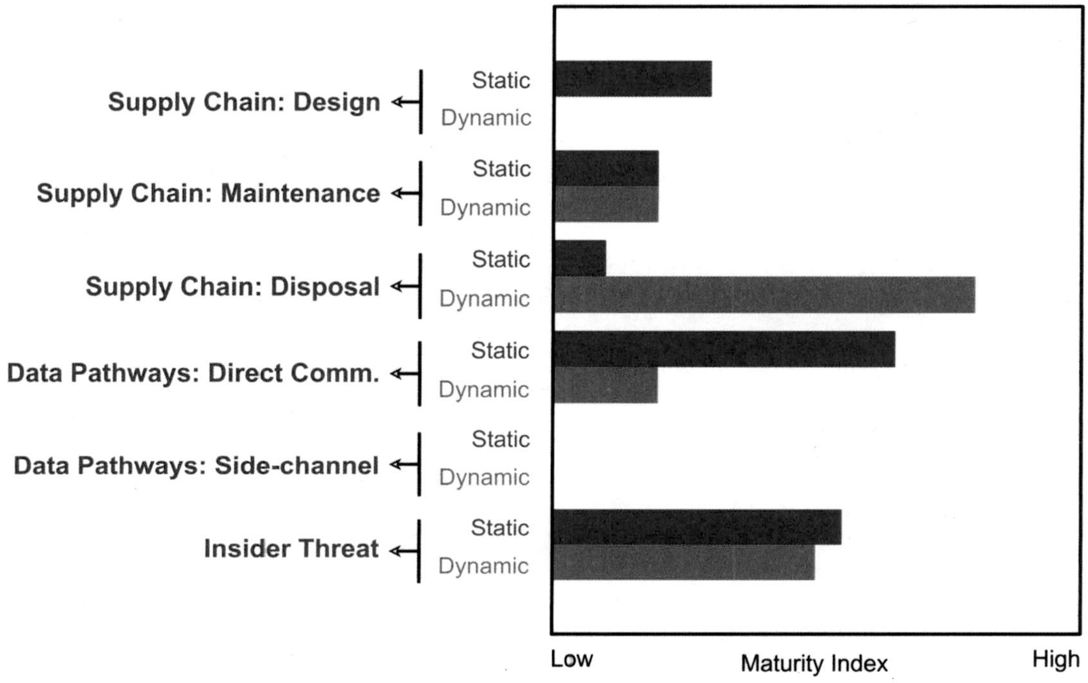

Assessing the Working-Level Assessments

No one can be entirely trusted to self-evaluate. Some other referee needs to assess the working-level assessments. For the acquisition community, three principal groups play this role for weapon systems: the program executive officers (PEOs) in the chain of command, authorizing officials exercising statutory authorities on behalf of the chief information officer, and developmental and operational testers. Each of these provides different checks of whether the program office has identified the appropriate countermeasures, implemented them appropriately, and assessed their effectiveness accurately.

Testing, for example, provides a fairly realistic assessment environment. However, even testing has its limitations. Most developmental testing focuses on the implementation of chosen solutions for the system under test rather than their effects on Red decisionmaking and actions. In operational testing, red team activities are most useful the broader ranging that they are. However, limitations of time and resources may limit the scope of operational testing. Ad hoc red-team exercises indicate as much about the skills of the red team as they do the vulnerabilities of the system under examination. The negative result that a red team fails to reveal significant vulnerabilities of a system only indicates that that team was unable to succeed in an attack or other compromise.[17] The results are only as good as the degree to which the red team skills mimic the real Red threat.

Each of these three evaluators must also be alert to shirking. Because no single actor is solely responsible for countermeasures for any given Red action, there is a potential tendency for some to shirk their responsibilities, assuming others are better positioned to address them. However, the key to cybersecurity and cyber resiliency is having multiple countermeasures against Red actions that combine to form an effective response. Even if the contribution of one actor might be small, it could still be important in the overall Blue response.

A bureaucratic difficulty with these assessments will be that many Red actions will have an immature set of Blue countermeasures. This circumstance might be satisfactory. Consider the case of an insider threat for someone handling nuclear weapons command and control. In this case, we might demand several Blue countermeasures, and high maturity for many of them, both in the static and dynamic dimensions. One example would be the personnel reliability program. If well implemented and tested, the maturity index would be high in this case. Now consider the case of an insider threat in an unclassified, noncritical system. A strict regime like the personnel reliability program would be unnecessarily burdensome. The Blue countermeasures overall should be less than in the nuclear case, and therefore a lower maturity score is fine. These critical systems can be identified by cyber mission thread analysis. The bureaucracy will need to adjust to the reality that not everything requires a high score.

A second factor that will play a role in determining how mature a set of Blue countermeasures to a Red action need to be is the threat. For Red actions of low threat, Blue might accept a lower maturity index; for Red actions in which the threat is high, the countermeasures need to have a comparably higher maturity.

Assessing the Mission

The third factor is the balance of countermeasures across the mission. Someone at the senior leader level needs to assess how well all of the countermeasures across all *systems* add up to an effective response for the cybersecurity and cyber resiliency of a *mission*. At this assessment level, a portfolio of working-level assessments, such as that shown in Figure 2.2, would be

---

[17] Stolfo, Bellovin, and Evans, 2011, p. 63.

overlain to get a holistic view of the susceptibility of the mission to cyber operations. At the mission level, are there gaps in Red coverage? Given the Boolean structure of the Red actions, are some successful countermeasures in one area offsetting poor performance in another (e.g., is leakage of information to Red about a system offset by a rapidly changing system configuration?)? Aggregating the view to a mission level in combination with cost and effectiveness information gives the perspective needed for resource allocation for cybersecurity and cyber resiliency.

Even if all of these assessments at the working level are done exquisitely, other issues can hobble Blue's efforts to thwart Red, and these lie at the institutional level of the enterprise, which we discuss in Chapter 3.

3. Monitoring at the Institutional Level

> "The harder we try to nail down quantitatively every detail on any given level, the more certain we are to overlook 'strategic' considerations at higher levels. One reason is that higher-level considerations . . . are inherently resistant to exact quantification. The other is that there are always broader perspectives from which a given problem can be viewed, and we cannot quantify them all."
>
> — James G. Roche and Barry D. Watts[1]

The Need for Monitoring Institutional Issues

In Chapter 2, we described a number of countermeasures that Blue can take in an attempt to thwart Red actions. If, after due diligence, Blue fails to hinder or quell Red's cyber operations, does that failure necessarily mean that the countermeasures were unsuccessful because of the people, ware, and the processes used, or could environmental factors in the institutions in which they are embedded have been the problem? The tendency of humans is to ascribe the blame for shortcomings to the people and objects closest in time and space to the failure—in this case the people, ware, and processes—rather than environmental factors, a tendency called the *fundamental attribution error*.[2] However, often the cause of a failure also has a strong component of institutional deficiencies. Knowing how well Blue is doing in cybersecurity and cyber resiliency requires going beyond monitoring people, ware, and processes. It also requires monitoring how well institutions are functioning.

The security of a mission or a system is not simply the sum of an assessment of the security of all of its elements. Security emerges from how all of these elements interact, how they mutually support or conflict with one another in achieving the goal of survivability and effectiveness. Just as a well-designed system is more reliable than its individual parts, the overall security of a system can exceed the security of its weakest elements or fall short of the average of its elements.[3] Metrics of military power, such as counting the number of divisions of ground forces and numbers of fighter aircraft and bombers, are insufficient for predicting the outcome of a conflict. Many factors, such as how those forces work together, the agility and robustness of their combat support, the stratagems and tactics they pursue, and how well they train, can also be

---

[1] Roche and Watts, 1991, p. 191.

[2] Ross, 1977; Langdridge and Butt, 2004. Humans present a range of other biases, many of which are discussed by Kahneman, 2011. We focus on the fundamental attribution error and, later in this chapter, drift, because these have been identified as critical to safety measures to avoid catastrophic accidents, which we argue has parallels to cybersecurity and cyber resiliency.

[3] See, for example, von Neumann, 1956.

decisive.[4] Because of the human element, cybersecurity and cyber resiliency have a strong social and organizational dimension. Feedback on the performance of institutions is critical for senior leaders in an enterprise to understand a mission's likely survivability and effectiveness when subjected to adversary cyber operations. Comprehensive security metrics should look less like project management tools and more like an insurance policy.[5]

Two very different case examples will help illustrate the fundamental attribution error in this context and point to the need for institutional-level cyber metrics. We first draw an example from the insider threat and the countermeasures used to mitigate against that vector.

Lessons from Poor Assessment of Human Risk

MAJ Nidal Malik Hasan was an Army psychiatrist who, over a period of years, performed poorly in his assigned duties, openly defended Osama bin Laden, justified suicide bombing, declared himself more loyal to Sharia law than to the U.S. Constitution repeatedly to his colleagues (orally and in written briefings), and was monitored by the Federal Bureau of Investigation (FBI) for plotting an attack with a known terrorist recruiter.[6] Hasan was an insider threat. The threat turned to action during a shooting rampage on November 5, 2009, in which he killed 13 and wounded 43 at Fort Hood, Texas.

By the principles outlined in Chapter 2, adequate mechanisms were in place to reveal Hasan as an insider threat. The Red action to be thwarted was clear: to prevent the recruitment of a malicious insider by a terrorist group. Blue countermeasures were in place, including a background check for a security clearance, policies and briefings to colleagues to report adverse or compromising information (for continuous monitoring), regular performance reviews, a disciplinary process, and counterintelligence activities, such as incidental FBI monitoring of his communications with a known terrorist recruiter. Why did all of these countermeasures fail?

The story is a long one with many details and nuances.[7] We review just a few factors here to illustrate the broader point. Although individuals did fail to perform their jobs as they ideally should, strong evidence indicates that the identity of the person in a given position did not play a large role. In other words, regardless of who the person was performing the countermeasure, the countermeasure failed and was likely to have failed even if different personnel were involved.

In the case of not disciplining Hasan in light of his poor job performance and other troubling actions, nine different individuals in his chain of command made the same mistake at various times.[8] To blame only the individuals involved would be a fundamental attribution error. This

---

[4] Hayward, 1968.

[5] Chapin and Akridge, 2005.

[6] Zegart, 2016.

[7] The story was thoroughly analyzed in Zegart, 2016.

[8] Zegart, 2016, p. 54.

pattern indicates an environmental factor—a failure of the institution—more so than the specific individuals involved. Some of those institutional factors are peculiarities of the situation at the time.

Psychiatrists, especially at his career stage, were in short supply in the Army during a period of high operational demands from post-traumatic stress disorder. Compounding this pressure were the normal disincentives for writing a strong, negative assessment that would lead to removal of an officer. If a leader recommends that someone be disciplined and puts the considerable effort into doing so, that leader had better be seen as correct by his or her superiors and peers. It is easier to pass the person along to the next assignment and let he or she worry about the individual. This factor was compounded in the case of Hasan because he was Muslim. The pressure was great to be correct in any negative assessment, and leaders were not well trained in discerning the difference between legitimate religious expression and dangerous radicalism. They were overly hesitant to judge his behavior.

Many coworkers also could have reported adverse behavior, as is encouraged for those holding security clearances. Surveys reveal that few workers report adverse behavior by their peers, often because they fear retribution, have a high threshold for reporting (only being willing to report clearly egregious behavior), and a distrust that the system will be fair to the subject of the reporting.[9] Peer reporting failed to identify Hasan, despite the large number of individuals who witnessed adverse behavior.

Finally, FBI counterintelligence-monitoring identified Hasan as a clear threat. However, two environmental factors stymied the investigation. First, institutional confusion among FBI field officers led to each thinking the other was on top of the case, whereas in reality neither was. Second, the agent assigned to investigate was a Department of Defense appointee in the FBI office who was selected for handling criminal cases and had little experience with counterterrorism. His investigation used poor methods, ones more appropriate for the fraud investigations he was trained to do than for counterterrorism. Individuals failed in their job performances, but other individuals also would likely have failed. The institutions had weak coordination, and individuals with inappropriate skills were placed in the key positions because of other institutional priorities.

These and other elements of the failure to detect Hasan as an insider threat point to deficiencies of risk assessment at the institutional level. That is not to say that individuals did not fail in this case and other cases. However, environmental factors clearly played a strong role. The institution did not always place the right person in the job, some personnel were insufficiently trained, and some felt that the organization had different priorities than leadership really held. The failure to detect and react to Hasan was not because Blue countermeasures were not in place—nor was it because personnel charged with carrying out those countermeasures were

---

[9] Wood, Crawford, and Lang, 2005.

trying not to do their jobs. It was, as Amy Zegart argues in some detail, primarily a failure of institutions.[10]

Addressing and monitoring only the working-level actions will be insufficient to detect these institutional-level factors and will leave decisionmakers with an incomplete picture of the state of cybersecurity and cyber resiliency. Weak institutional functioning can lead to failures even when workers are doing their best to accomplish the mission. On the other hand, strong institutions provide an important contributor to resiliency.

Lessons from Poor Assessment of Engineering Risk

Consider a second example, the well-studied events leading up to the loss of the Space Shuttle *Challenger* in 1986. The proximate cause of the loss of the vehicle and crew was the failure of an O-ring in one of the solid rocket boosters. However, the ultimate cause was an institutional failure. All of the engineering mechanisms to detect and address such concerns as the O-ring failure were in place throughout the program. No evidence exists that any workers were trying to shirk their responsibilities for safe design and operation of the vehicle. Feedback mechanisms at the working level looked fine. However, institutional-level deficiencies led to the acceptance of known engineering issues that eventually resulted in the loss of *Challenger*.[11]

Significant erosion of these seals, eventually including blow-by, had been observed in previous flights. Had these observations been made during developmental test and evaluation, it is unlikely that ignoring the engineering issues would have been considered acceptable risk. However, erosion incidents did not occur on every flight and had not yet led to complete failure. As Diane Vaughan, in her analysis of the *Challenger* launch decision, notes:

> Signals of potential danger lost their salience as a result of the risk assessment process. Accumulating incrementally, information about O-ring anomalies looked very different to the work group than it did [later] to outsiders, who viewed it knowing the disastrous outcome. Signals were mixed: information indicating trouble was interspersed with and/or followed by information signaling that all was well. Signals were weak: information was informal and/or ambiguous, so that the threat to flight safety was not clear. And, as 1985 drew to a close, signals were repeated, becoming routine as the frequency and predictability of erosion institutionalized the construction of risk.[12]

Larry Wear, one of the engineers involved with the O-ring flight certification at the National Aeronautics and Space Administration's (NASA) George C. Marshall Space Flight Center, expressed this concept rather bluntly:

> Once you've accepted an anomaly or something less than perfect, you know, you've given up your virginity. You can't go back. . . . I can imagine what the

---

[10] Zegart, 2016.

[11] For a full discussion, see Vaughan, 1996.

[12] Vaughan, 1996, pp. 243–244.

> Program would have said if we—we being me, Larry Mulloy, Thiokol, or anyone in between—if we had stood up one day and said, "We're not going to fly any more because we are seeing erosion on our seal." They would have looked back in the book and said, "Wait a minute. You've seen that before and you told us that was OK. And you saw it before that, and you said it was OK. Now what are you? Are you a wimp? Are you a liar? What are you?[13]

These kinds of deficiencies are deeply embedded in an institution. They reside in the perceptions of its members. What has occurred in these situations is that the organization's assessment of risk has drifted because the baseline shifted without realizing it until too late. *Risk*, in this case, is defined as the difference between the measured performance of some part of the enterprise and its desired performance.[14] It was not the working-level metrics that were at fault—the requisite information was within these organizations. It was the functioning of the institution that was at fault by embedding a false risk assessment, and the actors had become committed to this assessment. Working-level metrics do not capture this phenomenon.

There are instructive parallels between the institutional aspects of the assessment of the risk of O-ring failure in *Challenger* and the institutional aspects of assessment of cyber risk. Regarding the risk of blow-by of the solid-rocket-booster O-rings, the requisite engineering information was known within NASA. However, the signals that the safety margin was too small were intermittent and weak. No accident had yet occurred. The actors in the organization had assessed that the risk was acceptable, and it was hard to turn back on that assessment. It was difficult to say that the safety margin that was acceptable in the past was no longer acceptable and that *Challenger* should not fly.

Many cyber risks are also well known. The signals of exactly how systems or missions might be at risk are intermittent and often weak. A major mission loss, such as the crashing of an aircraft from a cyberattack, has not yet occurred. The Air Force has implicitly or explicitly assessed the risk of many of its systems and missions to be within acceptable security margins. Once that is done and those assessments are embedded in the culture, it is difficult to turn back and reassess the risk differently. Understanding the degree of such issues lies at the core of institutional metrics.

What to Monitor

It is all very well to note that, above and beyond working-level failings, institutional failings on behalf of Blue can be a significant cause of the inability to properly counter Red actions. However, what can be monitored in advance of a problem occurring that indicates how well Blue is performing at the institutional level? Is there a way to know that Blue countermeasures are at risk of failing to counter Red actions before Red acts?

---

[13] Interview with Larry Wear by Diane Vaughan, June 2, 1992, cited in Vaughan, 1996, p. 249.

[14] See also the discussion of defining risk in Fischhoff, Watson, and Hope, 1984.

An extensive literature, amassed from several decades of research, addresses this question.[15] The deductions emerge from two sets of observations: lessons extracted from catastrophic failures that have a strong institutional root cause and best practices of organizations that conduct complex operations successfully with minimal institutional failures.

Lessons from Failures

First, we look at some lessons from institutional failures. Institutional lapses that have led to catastrophic accidents have been well studied (e.g., the losses of the Space Shuttles *Challenger* and *Columbia*, the Three Mile Island and Chernobyl nuclear accidents, the Bhopal industrial accident). These provide valuable information for practices that can also lead to catastrophic security failures.

Any institution is characterized by the number, qualities, and interactions of all its personnel, equipment, and processes. We call this the *state* of the institution. The state includes how many manpower positions are authorized, the skills associated with them, and how many are filled, whether programs are fully or partially funded, whether units are at full readiness, the architecture of the command and control system, how many spares are available, the quality of policies, training, and so forth. In short, by *state*, we mean the totality of how the enterprise actually is and operates, not how it is measured or perceived. The state of the enterprise determines its cybersecurity and cyber resiliency.

A strong consensus is that major institutional malfunctions occur when fluctuations in the state of the enterprise exceed its ability to adapt to its environment (i.e., when fluctuations exceed its resiliency). Exceeding the tolerance envelope of the cyber state of an enterprise can occur because the fluctuations become too large, or because the ability of the institution to absorb these fluctuations atrophies. An example of the first would be if practices for Blue cyber hygiene fall behind Red's tradecraft. An example of the latter would be policies that excessively cut funding for cyber testing.

The antecedent to many institutional failures is a gradual closing of the gap between the fluctuations and the tolerance envelope of the institution's resiliency. Without senior leadership noticing, the margin of error and the resiliency of the enterprise erode. This phenomenon is called *drift*.[16] Drift can occur because the fluctuations are getting larger and eventually exceed the tolerance envelope, or because policies contract the size of the tolerance envelope until the fluctuations transgress it and resiliency is lost.

---

[15] Key readings, for example, include Perrow, 1999; Weick, 1987; Roberts, 1989; LaPorte, 1996; Sagan, 1993; Vaughan, 1996; Rasmussen, 1997; Dekker, 2006; Leveson et al., 2009; Shrivastava, Sonpar, and Pazzaglia, 2009; Winnefeld, Kirchhoff, and Upton, 2015.

[16] Betts, 1980–1981; Rasmussen, 1997; Dekker, 2006. For an extended case study, see Vaughan, 1996, where she calls *drift* "normalization of deviance."

What makes drift so insidious and difficult to discern is its gradual onset. Generally, nothing occurs abruptly; the changes that transpire go unnoticed by most members of the organization. In many cases when organizations experience a major malfunction, the cues warning of that failure existed somewhere in the organization prior to the failure.[17] Often, these signals were scattered at the working level but were not adequately assembled and directed to the right decisionmakers. Therefore, the problem tends to be less one of lack of working-level information; the problem is that the information is not adequately analyzed and directed to the right decision points.[18] Members of the organization in turn adjust their perceptions of risk accordingly. In the case of Hasan, all of the information to identify him as an insider threat existed within the government. The information was scattered, however, and was never assembled in the right way and delivered to the appropriate decisionmakers until it was too late. Also, all of the engineering issues in the *Challenger* loss existed within the enterprise. They, too, were not properly assessed and fed to the right decisionmakers. This dispersion of key working-level information leads to senior-level decisionmakers having perceptions of how operations work in an organization that depart from the reality of how they are really working.

A key indicator that drift is occurring is the appearance of a disconnect between how operations really work and how they are perceived to work by senior leaders.[19] A poignant example of this indicator of drift is the differences in the assessment of risk to vehicle and crew in the NASA Shuttle program. Richard Feynman (one of the members of the Presidential Commission investigating the *Challenger* accident) interviewed engineers at various levels within NASA, asking them how they assessed the probability of the catastrophic loss of a space shuttle. He observed that "it appears that there are enormous differences of opinion as to the probability of a failure with loss of vehicle and of human life. The estimates range from roughly 1 in 100 to 1 in 100,000. The higher figures come from the working engineers, and the very low figures from management."[20] Whether these numbers are accurate or not is less important than that they are so radically different—three orders of magnitude. Different levels in the organization of NASA had very different views of reality, an indication of institutional drift of risk assessment.

Because of the complexity of the cyber environment, disconnects between reality and how reality is perceived by senior leaders, or even mid-level leaders, can arise quite naturally. Organizations must work to avoid them. Leaders might believe that a network is isolated because it is air-gapped. However, they do not realize that a contractor connects his laptop to the network periodically during maintenance. A piece of maintenance equipment connected to a weapon system might receive software updates through unsecure channels, while leaders making risk

---

[17] Many researchers have made this observation. See, for example, Betts, 1980–1981.

[18] Simon, 1991.

[19] Freudenburg, 1988; Pidgeon and O'Leary, 2000; Hopkins, 2001; Dekker, 2006, especially pp. 82–86.

[20] Feynman, 1986.

assessments for the weapon system might be ignorant of this potential cross-flow of data. Information stored in a "cloud" might be exposed to an insider threat from a poorly vetted individual in the organization managing the cloud, unbeknownst to the client. Leaders might be unaware that the chain of custody of a software update might be variable and poorly controlled. The opportunities for such disconnects are countless. So, too, are the opportunities for drift. The risk of drift is so pronounced in the cyber realm that it is critical that monitoring and correcting these disconnects receive special attention.

Lessons from Success

The practices of organizations that avoid institutional failures over time while performing complex operations provide the keenest insight into what works best. Several such organizations have been studied, but an enterprise that stands out as an exemplar is U.S. commercial aviation. Commercial aviation has a remarkable record of operating thousands of flights per day under the management of numerous companies, in variable weather, with diverse aircraft designs and configurations, with only rare accidents. It was not always this way. Prior to the 1970s, accidents were far more common. The sharp decline in commercial aviation accidents in the United States is largely attributed to the creation of the Aviation Safety Reporting System (ASRS).[21]

The ASRS was established in 1975 to reduce the number of commercial airline accidents by sharing information that might mitigate accidents and making corrective recommendations to key decisionmakers. The ASRS is a self-correcting mechanism within the commercial aviation enterprise that enhances resiliency for safety. It is funded by the Federal Aviation Administration but implemented by NASA in order to be housed in an organization without regulatory authority. Aviation workers (pilots, air traffic controllers, dispatchers, flight attendants, maintenance technicians, and others) report incidents of concern to the ASRS. The reports are qualitative, voluntary, and legally protected as confidential.[22]

Subject-matter experts at the ASRS then analyze these data to look for trends and to identify latent problems. They can follow up on the reports as needed. When they discover issues of concern, they issue alert bulletins. The experience of these experts also facilitates the organization and indexing of the reports into a database that is useful for analysis by other experts interested in aviation safety. The ASRS is widely acclaimed as a successful program for feedback and learning and has been credited with playing a major role in the decline in commercial aircraft accidents since its inception.

The ASRS counters drift by providing vital feedback on institutional issues in a number of ways. First, the information can flow directly, without intermediate filters, from the operators closest to an issue through the ASRS to the highest-level officials in government and industry. In

---

[21] Dijkstra, 2006; Reynard et al., 1986; Vincent, 2003, pp. 198–199; Camm et al., 2013.

[22] Immunity is extended to reporters for all reports except those involving reportable accidents or criminal behavior (Reynard et al., 1986).

this way, the ASRS furnishes a channel for reporting of issues of concern outside of normal reporting chains, allowing information to flow to decisionmakers without risk of being blocked by the formal chain of command (which would be an example of an institutional deficiency).

Second, the reporting is voluntary, and most importantly, confidential. Confidentiality breeds trust, a necessary element for the level of probing necessary to uncover deep, systemic latent problems in an institution.[23] Confidentiality also provides feedback on the quality of safety and security not readily collected by any other means.[24] Voluntary reporting programs that empower lower-level participants who are protected by confidentiality also contribute to creating and sustaining a culture of safety within the enterprise.[25]

Third, meaning is created out of ASRS reported data by subject-matter experts. They examine incidents and areas of concern for possible latent consequences and also look for trends across organizations that might be missed because of organizational boundaries. The monitoring also looks for trends over time—specifically, drift. They can also do follow-up probing of issues of concern, to include further collection of data. Also, as outside experts, they provide an outsider view of issues, avoiding to some degree the bias that can arise from being a participant in a process.

Finally, the very existence of a mechanism like the ASRS helps foster a culture of safety or, in the context of this report, cybersecurity and cyber resiliency. The enterprise valuing the inputs of all of its members (regardless of position and stature), extending confidentiality, and analyzing, responding, and acting upon the information that it receives communicate forcefully to all members that safety (security) is of utmost value to the enterprise.

In these ways, the ASRS process facilitates closing the gap between how the aviation system actually operates and how it is perceived to operate, thereby helping to prevent drift and ensuring more resilient operations. It helps reveal if senior leaders perceiving operations as they really are.

In doing all of these activities, the ASRS feedback complements, but does not replace, working-level metrics. It is designed to capture and communicate the institutional deficiencies not reflected in more quantitative, routine monitoring of processes at the working level. It closes the gaps created by drift.

How to Monitor

Unlike the issues addressed by working-level cyber metrics, the issues addressed by institutional-level metrics do not arise from Red actions to counter. However, there are states of the institution to counter; the origin of these states is more internal than external. Chief among

---

[23] Pidgeon and O'Leary, 2000.

[24] O'Leary and Chappell, 1996.

[25] Pidgeon, 1997; Weick, 1987; Pidgeon and O'Leary, 2000.

these concerns is to reduce drift by ensuring that leaders' understanding of the state of the system does not depart significantly from reality. For this, Blue has some possible countermeasures.

Decisionmakers at all levels, but most critically at the highest levels, need to know whether the working-level feedback that they are receiving is missing anything or is being poorly assessed. The main manifestation of this kind of drift is when the perceptions of reality by senior leaders depart from reality in a way that leads to decisions that dangerously reduce institutional resiliency. The ultimate goal is to avoid the state of this disconnect. Institutional-level metrics are meant to reveal whether this goal is being realized.

To some extent, an organization can measure whether such disconnects exist by direct queries, much like the illustrative example of Richard Feynman polling Space Shuttle engineers and managers on their estimates of the probability of total loss of vehicle and crew. The problem with this approach is that senior leaders would need to know which questions to ask. To know which questions to ask presupposes that they have an idea where the problems lie.

Therefore, in this case, unlike the case of working-level metrics, the best that probably can be done is to monitor the qualities of the countermeasures to avoid the insidious, latent, institutional-level issues that can lead to security failures and, ultimately, ensure resiliency.

Fortunately, theory and practice indicate a sound set of countermeasures that reduce these disconnects. They are the practices instantiated by the ASRS. Specifically, the following questions reflect the maturity of mechanisms to monitor and reduce drift, in approximately increasing order of resiliency:

1. Do actors in the enterprise, regardless of their job or position, have mechanisms to report (a) any concerns they might have regarding the efficacy of Blue countermeasures against Red or (b) any ideas about how Blue countermeasures might be improved?
2. Can this feedback get to the right decisionmakers without being refracted by the priorities of the formal chain of command?
3. Does the potential reporter have any protections of confidentiality?
4. Is such reporting collected and assessed by subject-matter experts? Do they have the tools and skills to search for trends across organizations and over time?
5. Do the subject-matter experts have the ability to report their findings and recommendations to the appropriate decisionmakers without going through the chain of command?
6. Are these reporting and assessment processes reviewed periodically for their efficacy?

As a general rule, the deeper these questions can be answered in the affirmative, the better the feedback decisionmakers have on institutional-level drift, and the better the opportunities they have to make wise decisions that lead to enhanced cyber resiliency. Table 3.1 lists a proposed set of maturity levels for institutional-level issues.

Table 3.1. Maturity Levels for Institutional-Level Metrics

Level	Maturity	Characteristics
1	Highest maturity	Lessons are documented and distributed. Processes are reviewed periodically for improvement
2	Mature	Recommendations flow to relevant decisionmakers
3	Intermediate	Concerns are assessed by independent subject-matter experts and meaning is extracted from the reports in the form of recommendations
4	Immature	Formal mechanisms capturing key elements of the ASRS, including confidentiality, exist to report concerns regarding cybersecurity and cyber resiliency
5	Most immature	No mechanisms exist outside normal reporting through a chain of command for concerns that individuals have regarding cybersecurity and cyber resiliency

4. The Proper Use of Metrics

> "Whether information comes in a quantitative or qualitative flavor is not as important as how you use it."
>
> — Nate Silver[1]

We close with some general observations about implementing cyber metrics. It should be clear from the arguments made throughout this report that there is no simple "dashboard" of metrics for cybersecurity and cyber resiliency. No small set of easily observed qualities will reflect how well Blue is doing to ensure mission success in a cyber-contested environment. To assess Blue's performance for the cybersecurity and cyber resiliency of a system or mission, Blue needs to know how well it is curtailing the ability of Red to act against that system or mission. There are many ways that Red can act, and many of these ways do not lend themselves to quantitative assessment. Red's options will also evolve over time. So must Blue's responses. Because Red is adaptive and technologies evolve, what works well at one time with one technology might not work at a later time with a slightly different technology. Any use of metrics to evaluate how well Blue is doing must take these realities into account.

Humans interact with systems and manage missions. Because this human element is a central part of the cyber environment, cyber metrics must look over a much broader set of factors than, say, measuring how survivable and effective a weapon system will be in penetrating a particular air defense system. This breadth and complexity of cyber metrics point to the need for expertise and judgment in the workforce and among leaders and, therefore, introduces implementation challenges.[2]

Decisions

Levels

Because the purpose of metrics is to inform decisions, the selection of metrics should be guided by the decisions to be made, not by what is easily measured (or quantified). The types of metrics needed and the scope of those metrics depend on the decisionmaker being supported. No single set of metrics is well suited to all decisionmakers, and each measure is not necessarily useful for all decisionmakers.

---

[1] Silver, 2012, p. 72.

[2] The need for professional judgment in management, sometimes rather than metrics, is nearly ubiquitous (see Muller, 2018).

Technical decisions in development, production, and sustainment commonly are in most need of detailed, quantifiable metrics that tend toward the measures-of-performance end of the spectrum.

Operational decisions (e.g., for system requirements, concepts of operations, or operational planning) require output-oriented performance metrics, typically at a higher level of aggregation than used by the technical community. Metrics must be related to the threat and operational environment. These metrics can be a mix of both measures of performance and measures of effectiveness.

Strategic decisions often involve balancing the importance of the mission to service or national priorities with the perceived threat and available resources. Measures of effectiveness, particularly relative to other alternatives, are most useful.

Institutional decisions require measures of the true state of the organization and its processes. These metrics focus on the efficiency and effectiveness of the operation of the enterprise, with such questions as:

- What is the true state of our cybersecurity/cyber resilience posture?
- Is this being effectively assessed and communicated to leadership?
- Is the culture appropriate?
- How effective are the current cybersecurity processes?
- Are they actually being used?
- Do they achieve the desired results?
- How do we know?
- What are the mechanisms for feedback and are they used and effective?
- Are responsibility, authority, and accountability clearly assigned to appropriate officials?
- Are key elements of the enterprise properly resourced, and are personnel adequately trained?

In this report, we propose two sets of metrics: one set directed at the working-level and the second set directed at the institutional level of cybersecurity and cyber resiliency. Many decisionmakers and leaders will need to draw from the working-level metrics. However, we caution that only the highest-level leaders can make the decisions to change institutional culture, determine strategy, reallocate resources across the enterprise, establish new processes, and so on. It is vital that these high-level leaders get the institutional-level feedback outlined in Chapter 3. To focus on those issues, leaders at the top need to delegate responsibility for the working-level oversight to the appropriate levels. They need to hire and train good people, install sound processes, and empower them. Their central responsibility is to oversee the institutional aspects of the enterprise.

That is not to say that high-level leaders will never need to have working-level feedback. One example of this need is to monitor whether systemic issues are troubling workers across units. Are multiple units struggling with performance in the same working-level area? However, the focus of high-level leaders at the working level should be on systemic issues, not resolving specific working-level problems. Effective cyber countermeasures at the working level are nearly universally technical. The requisite information to enact strong, adaptive cyber countermeasures lies at the working-level, rarely at the highest levels of an organization. Leaders and decisionmakers at the highest levels should delegate those decisions to where the locus of information lies. Just as senior leaders delegate technical decisions to the working level in other domains, they need to allow the working level to select the right countermeasures to Red.

Goals

All decisionmakers need to keep their eye on the goal, which is operational mission assurance against Red cyber operational actions, not a set of proposed working-level Blue countermeasures. It is simply not feasible to list all actions that Blue needs to take to ensure the mission, both because of the complexity of the cyber domain and because of the evolving nature of the cyber threat.

However, the breadth of this goal is very challenging. One of those challenges is knowing whether enough is known to make sound decisions. At all levels, much of the feedback for cyber metrics is negative, which is to say, the absence of an issue. Some existing cyber vulnerabilities will not be discovered. Systems will not be successfully attacked until they are. Therefore, some of the feedback will leave decisionmakers unaware of real issues until those issues materialize. When something does *not* happen, for attacks that do *not* occur, what should decisionmakers take that to mean? Clearly, decisionmakers must be vigilant to avoid being lulled into complacency because they are ignorant of undiscovered vulnerabilities or that their systems have yet to be successfully attacked. Even if they have been attacked, the attacker might not have used his full capabilities.[3] Yet, decisionmakers cannot fret over all possible non-discovered problems and will have difficulties getting resources to address these issues without concrete evidence.

Not getting specific signifiers of a problem in the cyber realm until after an unwanted incident occurs, rather than getting many actionable sentinel alerts in advance, is common to the area of safety. Some lessons can be learned from this similarity. The predominate feedback on safety is that no accidents have occurred, then occasionally one does occur. Good managers are not lulled into complacency because no accidents have occurred. They maintain a culture of vigilance. They proactively seek indicators that safety measures are not being followed or that safety measures might not be adequate. They engage all members of the enterprise to actively identify any safety issues. They push against the reality that managers at all levels get rewarded for fixing problems; seldom do they get accolades for avoiding a problem that never happens.[4]

The implementation of cyber monitoring shares these characteristics with safety: Decisions are continuous, so monitoring must be continuous; the real state of cybersecurity and cyber resiliency spans the entire enterprise, so all members of the enterprise need to be involved in the

---

[3] See Snyder et al., 2015, pp. 15–16.

[4] See, for example, Repenning and Sterman, 2001.

monitoring and reporting of concerns; feedback will, nevertheless, be incomplete and sometimes inconsistent, so decisionmakers must interpret cyber metrics as part of a risk management process; and issues might be difficult to discover and the signals subtle, so decisionmakers must be proactive in probing the state of the enterprise.

Culture

Uncertainty

There are two kinds of uncertainty relevant to cyber metrics: uncertainty from random variations and uncertainty due to ignorance.[5] The first kind is often encountered in measurement and the ways of handling it statistically are well known. The latter is more problematic and has some intrinsic challenges that we highlighted in Chapter 3. Drift is one of the most insidious causes of ignorance and is a chronic problem in cyber metrics. It is easy to become inured by the status quo and to fail to adjust risk assessments in tune with the evolving technologies and threat environment. The only true way to know how well the enterprise is doing is to be attacked. Short of an attack, the most accurate information comes from intelligence (revealing the capabilities and intentions of Red) and developmental and operational testing (revealing some vulnerabilities). However, these data are still incomplete.

Decisionmakers need to accept these uncertainties and resist the temptation to press for inappropriate levels of precision and stability for working-level metrics when the environment is uncertain and dynamic. This uncertainty and lack of stability is why periodic assessment at the institutional level by knowledgeable personnel independent of the organization being assessed is so important. Accepting appropriate levels of risk is a key responsibility of senior leadership. Independent assessment helps ensure that these risk acceptance decisions are well informed.

This is another way of saying that cyber metrics are merely *indicators*. By themselves, cyber metrics rarely present a complete picture for decisionmaking. This is particularly true of MOEs because of their multidimensional nature. Decisionmakers need to keep in mind that the appropriate uses and limitations of cyber metrics must be realistically assessed and communicated; comparisons and trends should be examined and explained; and implications for the desired end state should be presented understandably.

Management

Cybersecurity and cyber resiliency are exercises in risk management. That means that some Blue countermeasures should have high maturity index scores if the Red threat is high and the consequences of not countering are also high. On the other hand, when the Red threat is low or the consequences of failure are less dire, accepting some risk might be appropriate. As developed

---

[5] See, for example, Ferson and Ginzburg, 1996.

in the framework for working-level metrics in Chapter 2, risk might also be accepted when weak performance in one area is compensated by stronger performance in another. Resources are always limited, and when managing, rather than eliminating risk, scoring low in selected areas is acceptable.

When a leader or unit is evaluated, there are strong pressures to score as high as possible. Chronically scoring low is hard to accept, either by those being evaluated or by outsiders wondering why performance is not higher. However, risk management necessarily accepts risk in some areas. So scores will be, and should be, low for some Red actions in some systems or mission elements. Accepting low scores, especially chronically low scores, will require some degree of a cultural change in the Air Force. Leaders will need to instill those changes.

People

Measures are only as good as the measurers. For cyber metrics, virtually every corner of the enterprise plays a role and, therefore, the measuring scope is vast. On top of that, it is fluid and requires deep reflection and insight—Blue needs to stay ahead of Red and inside of Reds' OODA loops. Blue personnel are the most critical resource in cyber monitoring. All personnel need to participate, and all need some level of training and skills. Because the observations are so often qualitative rather than quantitative, personnel must also *communicate* rather than just *report*. In many ways, defining the right cyber metrics is the easier, first step. The hard, next step is hiring, training, retaining, and keeping current a skilled workforce to execute those measures. Also, that workforce is not a subset of the personnel in the enterprise—as in safety, it is everyone in the enterprise.

References

Alpcan, Tansu, and Tamer Başar, *Network Security: A Decision and Game-Theoretical Approach*, New York: Cambridge University Press, 2011.

Anderson, Ross, *Security Engineering*, 2nd ed., Indianapolis, Ind.: Wiley, 2008.

Anonymous, "Can't Hack This," *New Scientist*, September 19, 2015, p. 20.

Arrow, Kenneth J., *The Limits of Organization*, New York: W.W. Norton & Company, 1974.

Bau, Jason, and John C. Mitchell, "Security Modeling and Analysis," *IEEE Security & Privacy*, Vol. 9, No. 3, May–June 2011, pp. 18–25.

Betts, Richard K., "Surprise Despite Warning: Why Sudden Attacks Succeed," *Political Science Quarterly*, Vol. 95, No. 4, Winter 1980–1981, pp. 551–572.

Boyens, Jon, Celia Paulsen, Rama Moorthy, and Nadya Bartol, *Supply Chain Risk Management Practices for Federal Information Systems and Organizations*, Washington, D.C.: U.S. Department of Commerce, NIST Special Publication 800-161, April 2015.

Bunn, Matthew, and Scott D. Sagan, "A Worst Practices Guide to Insider Threats," in Matthew Bunn and Scott D. Sagan, eds., *Insider Threats*, Ithaca, N.Y.: Cornell University Press, 2016, pp. 145–174.

Camm, Frank, Laura Werber, Julie Kim, Elizabeth Wilke, and Rena Rudavsky, *Charting the Course for a New Air Force Inspection System*, Santa Monica, Calif.: RAND Corporation, TR-1291-AF, 2013. As of August 28, 2018:
https://www.rand.org/pubs/technical_reports/TR1291.html

Cappelli, Dawn, Andrew Moore, and Randall Trzeciak, *The CERT® Guide to Insider Threats*, Upper Saddle River, N. J.: Addison-Wesley, 2012.

CERT Insider Threat Center, *Common Sense Guide to Mitigating Insider Threats*, 5th ed., Pittsburgh, Penn.: Carnegie Mellon University, Software Engineering Institute Technical Note CMU/SEI-2016-TR-015, December 2016.

Chapin, David A., and Steven Akridge, "How Can Security Be Measured?" *Information Systems Control Journal*, Vol. 2, 2005, pp. 43–47.

Cheng, Yi, Julia Deng, Jason Li, Scott DeLoach, Anoop Singhal, and Xinming Ou, "Metrics of Security," in Alexander Kott, Cliff Wang, and Robert F. Erbacher, eds., *Cyber Defense and Situational Awareness*, Advances in Information Security, Vol. 62, Springer, 2014, pp. 263–295.

Churchman, C. West, and Russell L. Ackoff, "An Approximate Measure of Value," *Journal of the Operations Research Society of America*, Vol. 2, No. 2, May 1954, pp. 172–187.

Committee on National Security Systems, *Security Categorization and Control Selection for National Security Systems*, Fort Meade, Md.: Committee on National Security Systems, CNSSI No. 1253, March 27, 2014.

Defense Acquisition University, *Glossary of Defense Acquisition Acronyms and Terms*, 16th edition, Fort Belvoir, Va.: Defense Acquisition University Press, September 2015.

Degabriele, Jean Paul, G. Paterson, and Gaven J. Watson, "Provable Security in the Real World," *IEEE Security & Privacy*, Vol. 9, No. 3, May–June 2011, pp. 33–41.

Dekker, Sidney, "Resilience Engineering: Chronicling the Emergence of Confused Consensus," in Erik Hollnagel, David D. Woods, and Nancy Leveson, eds., *Resilience Engineering: Concepts and Precepts*, Burlington, Vt.: Ashgate, 2006, pp. 76–92.

Department of Defense Instruction 5000.61, *DoD Modeling and Simulation (M&S) Verification, Validation, and Accreditation (VV&A)*, Washington, D.C.: U.S. Department of Defense, December 9, 2009.

Dijkstra, Arthur, "Safety Management in Airlines," in Erik Hollnagel, David D. Woods, and Nancy Leveson, eds., *Resilience Engineering: Concepts and Precepts*, Burlington, Vt.: Ashgate, 2006, pp. 183–203.

Ferson, Scott, and Lev R. Ginzburg, "Different Methods Are Needed to Propagate Ignorance and Variability," *Reliability Engineering and System Safety*, Vol. 54, No. 2–3, 1996, pp. 133–144.

Feynman, R. P., "Volume 2, Appendix F: Personal Observations on Reliability of Shuttle," in *Report of the Presidential Commission on the Space Shuttle Challenger Accident*, Washington, D.C., 1986.

Fischhoff, Baruch, Stephen R. Watson, and Chris Hope, "Defining Risk," *Policy Sciences*, Vol. 17, 1984, pp. 123–139.

Freedman, Lawrence, *Strategy: A History*, New York: Oxford University Press, 2013.

Freudenburg, William R., "Perceived Risk, Real Risk: Social Science and the Art of Probabilistic Risk Assessment," *Science*, Vol. 242, No. 4875, 1988, pp. 44–49.

Hayward, Philip, "The Measurement of Combat Effectiveness," *Operations Research*, Vol. 16, No. 2, 1968, pp. 314–323.

Hitch, Charles J., and Roland N. McKean, *The Economics of Defense in the Nuclear Age*, Cambridge, Mass.: Harvard University Press, 1960.

Hopkins, Andrew, "Was Three Mile Island a 'Normal Accident'?" *Journal of Contingencies and Crisis Management*, Vol. 9, No. 2, 2001, pp. 65–72.

Howard, Michael, Jon Pincus, and Jeannette M. Wing, "Measuring Relative Attack Surfaces," in D. T. Lee, S. P. Shieh, and Doug Tygar, eds., *Computer Security in the 21st Century*, Springer, 2005, pp. 109–137.

Hubbard, Douglas W., and Richard Seieren, *How to Measure Anything in Cybersecurity Risk*, Hoboken, N.J.: John Wiley & Sons, 2016.

Hvistendahl, Mara, "3D Printers Vulnerable to Spying: Design Information Can Be Pilfered from the Sounds a Printer Makes," *Science*, Vol. 352, No. 6282, April 8, 2016, pp. 132–133.

James, William, "The Will to Believe," *The New World: A Quarterly Review of Religion, Ethics and Theology*, Vol. 5, June 1896, pp. 327–347.

Jansen, Wayne, *Directions in Security Metrics Research*, Washington, D.C.: National Institute of Standards and Technology, NISTIR 7564, April 2009.

JASON, *Science of Cyber-Security*, McLean, Va.: MITRE Corporation, JSR-10-102, November 2010.

JCIDS—*See* Joint Capabilities Integration and Development System.

Jensen, Michael C., and William H. Meckling, "Specific and General Knowledge, and Organizational Structure," in *Contract Economics*, L. Werin and H. Hijkander, eds., Cambridge, Mass.: Basil Blackwell, 1992, pp. 251–274.

Joint Capabilities Integration and Development System, *Manual for the Operation of the Joint Capabilities Integration and Development System (JCIDS)*, February 12, 2015, including errata as of December 18, 2015.

Kahneman, Daniel, *Thinking, Fast and Slow*, New York: Farrar, Straus and Giroux, 2011.

Kallberg, Jan, and Thomas S. Cook, "The Unfitness of Traditional Military Thinking in Cyber: Four Cyber Tenets That Undermine Conventional Strategies," *IEEE Access*, Vol. 5, 2017, pp. 8126–8130.

Langdridge, Darren, and Trevor Butt, "The Fundamental Attribution Error: A Phenomenological Critique," *British Journal of Social Psychology*, Vol. 43, No. 3, September 2004, pp. 357–369.

LaPorte, Todd R., "High Reliability Organizations: Unlikely, Demanding and at Risk," *Journal of Contingencies and Crisis Management*, Vol. 4, No. 2, June 1996, pp. 60–71.

Leveson, Nancy, Nicolas Dulac, Karen Marais, and John Carroll, "Moving Beyond Normal Accidents and High Reliability Organizations: A Systems Approach to Safety in Complex Systems," *Organization Studies*, Vol. 30, No. 2–3, 2009, pp. 227–249.

Marshall, A. W., *Problems of Estimating Military Power*, Santa Monica, Calif.: RAND Corporation, P-3417, August 1966. As of August 28, 2018: https://www.rand.org/pubs/papers/P3417.html

Mitchell, Robert, and Ing-Ray Chen, "A Survey of Intrusion Detection Techniques for Cyber-Physical Systems," *ACM Computing Surveys*, Vol. 46, No. 4, Article 55, March 2014.

Muller, Jerry Z., *The Tyranny of Metrics*, Princeton, N.J.: Princeton University Press, 2018.

National Institute of Standards and Technology, *Security and Privacy Controls for Federal Information Systems and Organizations*, NIST Special Publication 800-53, Revision 4, Washington, D.C.: U.S. Department of Commerce, last updated January 2015.

Office of Aerospace Studies, *The Measures Handbook*, Kirtland Air Force Base, N.M.: Air Force Materiel Command, August 6, 2014.

O'Leary, Mike, and Sheryl L. Chappell, "Confidential Incident Reporting Systems Create Vital Awareness of Safety Problems," *ICAO Journal*, Vol. 51, No. 8, 1996, pp. 11–13.

Perrow, Charles, *Normal Accidents: Living with High-Risk Technologies*, Princeton, N.J.: Princeton University Press, 1999.

Pfleeger, Shari Lawrence, "Useful Cybersecurity Metrics," *IT Professional*, Issue 3, May–June 2009, pp. 38–45.

———, "Security Measurement Steps, Missteps, and Next Steps," *IEEE Security & Privacy*, Vol. 10, No. 4, July–August 2012, pp. 5–9.

Pfleeger, Shari Lawrence, and Robert K. Cunningham, "Why Measuring Security Is Hard," *IEEE Security & Privacy*, Vol. 8, No. 4, July–August 2010, pp. 46–54.

Pidgeon, Nick, "The Limits to Safety? Culture, Politics, Learning and Man-made Disasters," *Journal of Contingencies and Crisis Management*, Vol. 5, No. 1, 1997, pp. 1–14.

Pidgeon, N., and M. O'Leary, "Man-Made Disasters: Why Technology and Organizations (Sometimes) Fail," *Safety Science*, Vol. 34, No. 1–3, 2000, pp. 15–30.

Rasmussen, Jens, "Risk Management in a Dynamic Society: A Modelling Problem," *Safety Science*, Vol. 27, No. 2–3, 1997, pp. 183–213.

Repenning, Nelson P., and John D. Sterman, "Nobody Ever Gets Credit for Fixing Problems That Never Happened: Creating and Sustaining Process Improvement," *California Management Review*, Vol. 43, No. 4, Summer 2001, pp. 64–87.

Reynard, W. D., C. E. Billings, E. S. Cheaney, and R. Hardy, *The Development of the NASA Aviation Safety Reporting System*, NASA Reference Publication 1114, November 1986.

Roberts, Karlene H., "New Challenges in Organizational Research: High Reliability Organizations," *Industrial Crisis Quarterly*, Vol. 3, No. 2, 1989, pp. 111–125.

Roche, James G., and Barry D. Watts, "Choosing Analytic Measures," *Journal of Strategic Studies*, Vol. 14, No. 2, 1991, pp. 165–209.

Ross, Lee, "The Intuitive Psychologist and His Shortcomings: Distortions in the Attribution Process," in L. Berkowitz, ed., *Advances in Experimental Social Psychology*, Vol. 10, New York: Academic Press, 1977, pp. 173–220.

Sagan, Scott D., *The Limits of Safety: Organizations, Accidents, and Nuclear Weapons*, Princeton, N.J.: Princeton University Press, 1993.

Scarfone, Karen, and Peter Mell, *Guide to Intrusion Detection and Prevention Systems (IDPS)*, Gaithersburg, Md.: National Institute of Standards and Technology, Special Publication 800-94, February 2007.

Sedgewick, Adam, Murugiah Souppaya, and Karen Scarfone, *Guide to Application Whitelisting*, Washington, D.C.: U.S. Department of Commerce, NIST Special Publication 800-167, October 2015.

Shackleford, Dave, *Combatting Cyber Risks in the Supply Chain*, SANS Institute, September 2015.

Shi, Qingkai, Zhenyu Chen, Chunrong Fang, Yang Feng, and Baowen Xu, "Measuring the Diversity of a Test Set with Distance Entropy," *IEEE Transactions on Reliability*, Vol. 65, No. 1, March 2016, pp. 19–27.

Shin, Yonghee, Andrew Meneely, Laurie Williams, and Jason A. Osborne, "Evaluating Complexity, Code Churn, and Developer Activity Metrics as Indicators of Software Vulnerabilities," *IEEE Transactions on Software Engineering*, Vol. 37, No. 6, November–December 2011, pp. 772–787.

Shrivastava, Samir, Karan Sonpar, and Federica Pazzaglia, "Normal Accident Theory Versus High Reliability Theory: A Resolution and Call for an Open Systems View of Accidents," *Human Relations*, Vol. 62, No. 9, 2009, pp. 1357–1390.

Silver, Nate, *The Signal and the Noise: Why So Many Predictions Fail—But Some Don't*, New York: Penguin Press, 2012.

Simon, Herbert A., "Bounded Rationality and Organizational Learning," *Organization Science*, Vol. 2, No. 1, 1991, pp. 125–134.

Snyder, Don, James D. Powers, Elizabeth Bodine-Baron, Bernard Fox, Lauren Kendrick, and Michael H. Powell, *Improving the Cybersecurity of U.S. Air Force Military Systems Throughout Their Life Cycles*, Santa Monica, Calif.: RAND Corporation, RR-1007-AF,

2015. As of August 28, 2018:
https://www.rand.org/pubs/research_reports/RR1007.html

Stolfo, Sal, Steven M. Bellovin, and David Evans, "Measuring Security," *IEEE Security & Privacy*, May–June 2011, pp. 60–65.

Stoneburner, Gary, *Underlying Technical Models for Information Technology Security: Recommendations of the National Institute of Standards and Technology*, Gaithersburg, Md.: National Institute of Standards and Technology, NIST Special Publication 800-33, December 2001.

U.S. Department of Defense, *DOD Dictionary of Military and Associated Terms*, Washington, D.C. July 2017.

U.S. Department of Energy and U.S. Department of Homeland Security, *Cybersecurity Capability Maturity Model (C2M2)*, Version 1.1, February 2014.

U.S. Department of Homeland Security and Executive Office of the President of the United States, *FY 2016 CIO FISMA Metrics*, Version 1.0, October 2015.

Vaughan, Diane, *The Challenger Launch Decision: Risky Technology, Culture, and Deviance at NASA*, Chicago: University of Chicago Press, 1996.

Vicente, Kim, *The Human Factor*, New York: Routledge, 2003.

von Neumann, J., "Probabilistic Logics and the Synthesis of Reliable Organisms from Unreliable Components," in C. E. Shannon and J. McCarthy, eds., *Automata Studies*, Princeton University Press, 1956, pp. 43–98.

Wang, Cliff, and Zhuo Lu, "Cyber Deception: Overview and the Road Ahead," *IEEE Security & Privacy*, March–April 2018, pp. 80–85.

Wang, Lingyu, Sushil Jajodia, Anoop Singhal, Pengsu Cheng, and Steven Noel, "*k*-Zero Day Safety: A Network Security Metric for Measuring the Risk of Unknown Vulnerabilities," *IEEE Transactions on Dependable and Secure Computing*, Vol. 11, No. 1, January–February 2014, pp. 30–44.

Warwick, Graham, "Secure by Design: Mathematically Exact Software Analysis and Design Reduce Cybervulnerability in Flight Tests," *Aviation Week & Space Technology*, November 23–December 6, 2015, pp. 69–70.

Weick, Karl E., "Organizational Culture as a Source of High Reliability," *California Management Review*, Vol. 24, No. 2, Winter 1987, pp. 112–127.

Winnefeld, James A., "Sandy," Jr., Christopher Kirchhoff, and David M. Upton, "Cybersecurity's Human Factor: Lessons from the Pentagon," *Harvard Business Review*, September 2015, pp. 86–95.

Wood, Suzanne, Kent S. Crawford, and Eric L. Lang, *Reporting of Counterintelligence and Security Indicators by Supervisors and Coworkers*, Monterey, Calif.: Defense Personnel Security Research Center, Technical Report 05-6, May 2005.

Yee, George O. M., "Security Metrics: An Introduction and Literature Review," in John R. Vacca, ed., *Computer and Information Security Handbook*, Waltham, Mass.: Morgan Kaufmann Publishers, 2013, pp. 553–566.

Zegart, Amy B., "The Fort Hood Terrorist Attack: An Organizational Postmortem of Army and FBI Deficiencies," in Matthew Bunn and Scott D. Sagan, eds., *Insider Threats*, Ithaca, N.Y.: Cornell University Press, 2016, pp. 42–73.

Zhang, Mengyuan, Lingyu Wang, Sushil Jajodia, Anoop Singhal, and Massimiliano Albanese, "Network Diversity: A Security Metric for Evaluating the Resilience of Networks Against Zero-Day Attacks," *IEEE Transactions on Information Forensics and Security*, Vol. 11, No. 5, May 2016, pp. 1071–1086.